Model Archiving and Sustainment for Aerospace Design

Model Archiving and Sustainment for Aerospace Design

SEAN BARKER

SAE INTERNATIONAL®

Warrendale, Pennsylvania, USA

SAE INTERNATIONAL®

400 Commonwealth Drive
Warrendale, PA 15096-0001 USA
E-mail: CustomerService@sae.org
Phone: +1877-606-7323 (inside USA and Canada)
           +1724-776-4970 (outside USA)
FAX: +1724-776-0790

Library of Congress Catalog Number 2020930074
http://dx.doi.org/10.4271/9781468601336

Information contained in this work has been obtained by SAE International from sources believed to be reliable. However, neither SAE International nor its authors guarantee the accuracy or completeness of any information published herein and neither SAE International nor its authors shall be responsible for any errors, omissions, or damages arising out of use of this information. This work is published with the understanding that SAE International and its authors are supplying information, but are not attempting to render engineering or other professional services. If such services are required, the assistance of an appropriate professional should be sought.

ISBN-Print 978-1-4686-0132-9

To purchase bulk quantities, please contact: SAE Customer Service

E-mail: CustomerService@sae.org
Phone: +1877-606-7323 (inside USA and Canada)
           +1724-776-4970 (outside USA)
Fax: +1724-776-0790

Visit the SAE International Bookstore at books.sae.org

Chief Product Officer
Frank Menchaca

Publisher
Sherry Dickinson Nigam

Director of Content Management
Kelli Zilko

Production Associate
Erin Mendicino

Manufacturing Associate
Adam Goebel

contents

CHAPTER 6

Governance, Planning, and Preservation Planning   51

CHAPTER 7

Basics of CAD   59

CHAPTER 8

Preserving CAD   73

CHAPTER 9

Signposts for Other Models   81

CHAPTER 10

The Basics of PDM   89

CHAPTER 11

How to Archive an Aircraft   101

CHAPTER 12

Summary and Future Directions   109

acknowledgments

My thanks goes first to all the people who wrote the standards and reports I have cited, including OAIS, TRAC, STEP, EIA 649, and ISO 14926. This book also redistributes many concepts harvested from conversations with colleagues, and there are many who have supported work in other ways. The following deserve particular mention:

- **Jean-Yves Delauney** and **Rick Zuray** for leading LOTAR so successfully for so long

- The participants in LOTAR, and supporting organizations, particularly PDES and ProSTEP

- Josh Lubel of NIST and Professor Chris McMahon for projects such as KIM

- RASCC program colleagues, especially Matthew Addis, Jonathan Cave, and Rob Howard

- Phil Spiby for background from the pharmaceutical industry

- Alex Ball for his work on CAD archiving

- William Kilbride and the Digital Preservation Coalition for promoting understanding of the subject

- Brian Yarrow of BAE Systems who suggested I work on LOTAR and Howard Mason for funding it

- BSI ACE/1 committee for supporting UK involvement in LOTAR

- Professor Alan Middleditch who introduced me to CAD software design

- And—of course—Wikipedia, for providing so many accessible summaries of the sources

Now notionally retired, Barker worked for BAE Systems for 30 years—20 for the Advanced Technology Centre after an initial 10 years for the Warton aircraft factory, the latter including a 3-year secondment with Dassault Aviation in Paris. He previously worked 7 years for Marconi Avionics and 3 in academia.

In academia, he worked on CAD from the inside, developing algorithms for biarc curves and software supporting constructive solid geometry. This led to his running the CAD group at BAE Systems, developing their surface modeler. An internal move saw him running the Product Data Management group, where, acting as PDM architect, he helped develop concepts for model-based PDM and contributed to the product structure data exchange between the Eurofighter partners. During this period he also contributed to the use of the STEP standard in BAE Systems.

The PDM work continued through the secondment in Paris, whence he moved to the ATC in Bristol in 2000. There he became again involved in STEP, including as a data modeler for Product Life Cycle Support (PLCS) standard. He joined the LOTAR team in 2002, contributing to parts on governance and PDM, and editing the glossary. The work at the ATC was wide ranging and included projects on design integration and service architectures, as well as other STEP and archiving projects.

Other roles included BAE Systems technical advisory board member at PDES Inc., UK technical expert for the British Standards Institute on long-term archiving of aircraft data, vice chair of the European Workshop Agreement on the Tactical Situation Object (for emergency response), and a member of the British Computing Society academic awards committee. His first degree was in Mathematics from Imperial College London, and he is a chartered engineer and fellow of the British Computing Society.

Introducing Data Retention: Why? What? and How?

1.1 Why?: Data Sustainment and Business Risk

A plane crashes, people die, and the aviation authorities investigate. The aircraft manufacturer—the aircraft's design authority—is asked to supply the design, along with supporting information that shows the designers were thorough and diligent (and hence not negligent), so that investigators can find the cause. A story—probably apocryphal—has a senior aircraft executive being shown a huge hangar full of filing cabinets and being told that this is where the design documents are stored, but the guys who remember the filing system have long since retired. At least the executive can hire lawyers (at several hundred dollars an hour) to sift through.

Update that story for the digital age, rather than having a hangar full of filing cabinets, the executive would find racks holding several different types of tape, a graveyard of disk drives, and a tower of CD ROMs. But it's not just the guys who knew the filing system who have retired, so have the computer operators who know which computer to hook the drives up to, assuming that type of computer can be borrowed from a museum. And then they will need to match the tapes to the software that can make sense of them and hope the company that reissues the software license keys is still in business. But, although this is a problem with IT, it is not an IT problem. An IT supplier will sell you the hardware or lease the software, but very rarely will it take responsibility for your data. Data retention is a problem of business planning, particularly of ensuring that the plan to sustain the business is followed through all the way to retaining important information.

Fortunately, this updated story is a work of imagination, and the aerospace industry has been working in earnest on digital archiving (or retention or sustainment) since the early 2000s. This book provides a gentle introduction to the core of that work. A key message is that digital retention is 95% a business problem, and only 5% a problem that the IT suppliers will help with.

That message starts with the three main business reasons for sustaining data. The first is regulatory compliance: in the aerospace sector, an aircraft may not fly unless the aircraft design is certified as airworthy, and this means the regulators must be able to access the design data throughout the operational life of the aircraft. Other sectors have other requirements, for example, the pharmaceutical industry must retain drug trial data, and many countries have laws such as the US Sarbanes–Oxley Act [1.1] on keeping corporate accounting and finance records. The second reason is to provide evidence in litigation, particularly with regard to product liability, and showing, for example, that a product complies with regulations and safety standards. The third is design reuse, typically avoiding the cost of recreating a lost original. An additional, emerging requirement is to mine operational data, with the aims of reducing maintenance costs and optimizing aircraft operations.

So, the people who should read the first chapter are the shareholders and senior executives of aerospace companies. It is in the company's interest that digital retention is properly understood and resourced. Without long-term data retention, the long-term profitability of the company is at risk. Engineering managers and senior technologists should read the first half of the book, which covers the general concepts of archiving. They need to understand what is required, since they will have to implement the solution. And beyond that, much of the detail will be delegated to professional engineers and technologists, for whom the second half goes into some technical detail. This book is not "the solution to archiving CAD," rather it describes a general methodology for archiving all types of design model, but using computer-aided design (CAD) as a detailed example.

Beyond the aerospace industry, IT outsourcing and software companies should understand the concerns of their customers—and need reminding that their customers' motivation is not to buy more IT but ensure their business keeps making a profit. And the relevant customers will include the nuclear industry, oil refining and chemical engineering industries, and the automotive and ship building industries—in fact, any industry producing an engineered product that has a lifecycle of 10 or more years.

1.2 The Sustainment Problem and Why It Is Coming to the Forefront Now

The basics: IT-based business processes involve a combination of data and software. The data is stored at fixed points in the process, rather like fence posts. The software takes the data from one fixed point and transforms it into the data needed at the next fixed point—the wire between the posts. To make a fence, you need both wire and fence posts—both the data and the software.

Most software contains bugs, and bugs in new versions of software can distort designs made under the previous versions. In effect, they take the old design data, but they produce a different design from the previous version of the software. Figure 1.1 shows what happened when a part designed in one generation of mechanical CAD software was first imported into the next generation [1.2]. The result is clearly different and could not fit in the same place as the original. This at least is a blatant error. Subtler errors might not be noticed, but could invalidate the airworthiness analysis and might even lead to a crash.

FIGURE 1.1 Errors arising from software generation change.

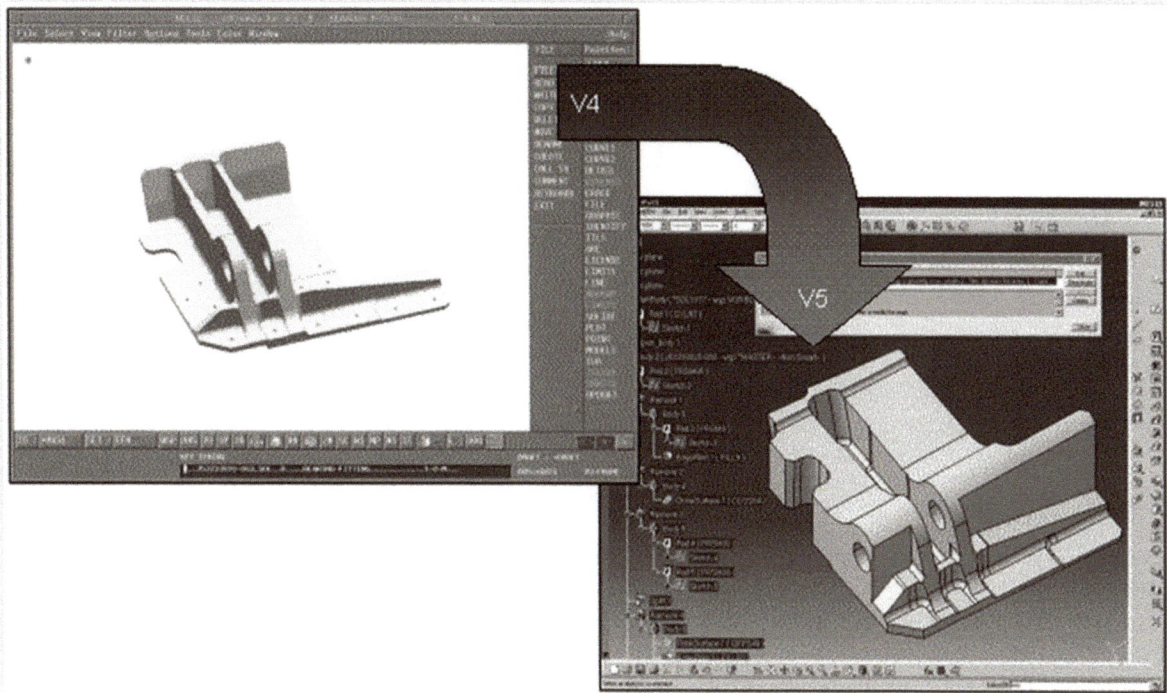

In older, document-based processes, the output of a design process was a drawing—whether designed on a computer or not—and it was the physical drawing that was approved. People took measurements from the drawing as the inputs for the downstream process. To make a new copy of the drawing it was photocopied (blueprinting was a pre-Xerox photocopying process). Key here is that you don't need to rerun the design software to access the drawing, and so the drawing does not change as new software is introduced.

The problem for the aircraft industry is that the lifecycle of an aircraft design is of the order of 70 years. This is much longer than the lifecycle of a CAD system, which, when this work started in earnest in the 2000s, was about 6 months between updates and 10 years between generations (major software rewrites)—although it seems now that older generations of software are being maintained in parallel with newer generations, probably because the users do not want the disruption that a new software generation would cause.

Moreover, most software relies on operating system functions and services. The computers and their operating systems used to change every 3 years, creating a further route for errors to creep in. Figure 1.2 [1.2] shows this lifecycle clash: it is based on a figure first used in the early 2000s, revised to show 20 years between CAD generations. The top two lines show the CAD system lifecycle, and the lower lines show lifecycle of various aircraft revenue streams. Assuming design starts in the first quarter of a CAD generation, then CAD system obsolescence becomes an issue in the latter phases of the production run, which are still the early phases of the various in-service revenue streams.

Why is this problem coming to the forefront? Actually, the aerospace industry has been well aware of incompatibilities between software versions since the 2000s, which added $6 billion to the development costs of the Airbus A380 [1.3]. The aircraft industry started adopting model-based CAD in the 1990s, and they had a history of investing in CAD long before the processes became model-based. However, since the 2000s, model-based processes have extended beyond the design of metal parts to include electrical

FIGURE 1.2   Lifecycle clash.

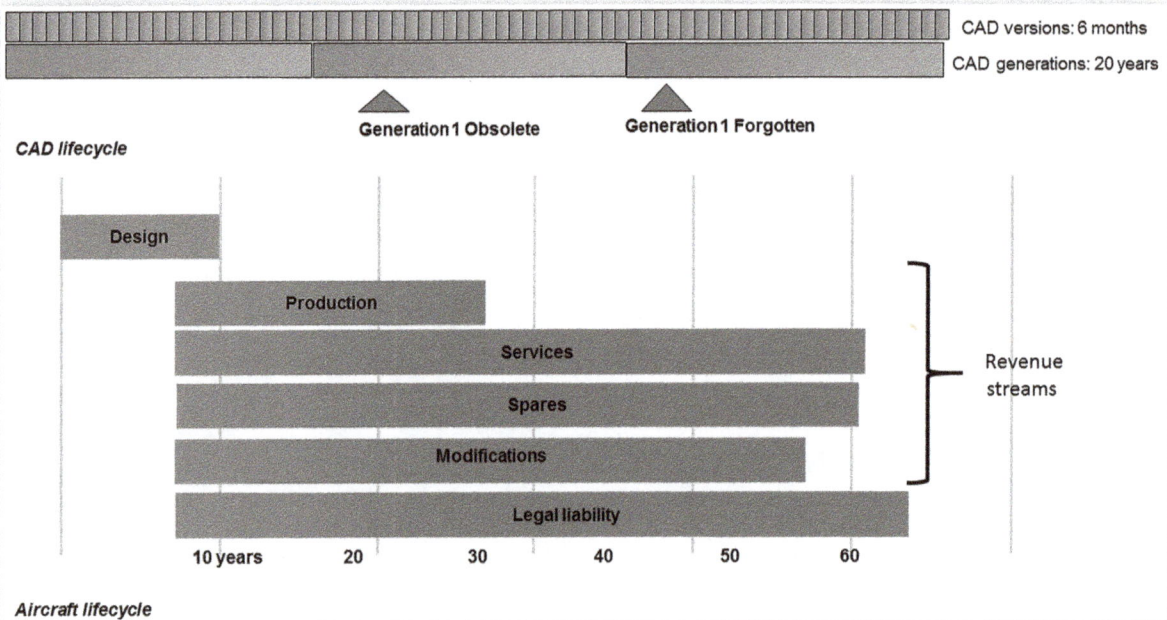

	CAD versions: 6 months
	CAD generations: 20 years

Generation 1 Obsolete   Generation 1 Forgotten

CAD lifecycle

Design

Production

Services

Spares   } Revenue streams

Modifications

Legal liability

10 years   20   30   40   50   60

Aircraft lifecycle

systems, composite parts, systems engineering, and even requirements engineering. And, with the increasing availability of model-based software, other industries are adopting the software to remain competitive and have now been doing it for long enough to start to recognize the problems of long-term sustainment.

1.3 Retain What? Knowledge, Information, or Data?

An item of data is a value, for example, the number 200. It represents information—a fact about a system—for instance, the number 200 could be a readout from an altimeter showing the height of an aircraft. Knowledge involves the intelligent use of information, for example, knowing that as you come out of a low-level run over the sea, 200 ft is too low to fly over land and you need to climb.

On this model, long-term *data* retention involves ensuring the bits and bytes of data can be recovered uncorrupted from an archive. Here a key topic is media migration—ensuring that the floppy disks of yesteryear are copied before the magnetic media decays or there are no computers left with floppy drives to read the disks.

Information retention involves not only keeping the data available and uncorrupted, but also guaranteeing that the software reads the data correctly. Figure 1.1 showed an error caused by changes in software, not changes in the data.

Information is a snapshot of its context, and that context links to a broad network of auxiliary knowledge that aids in its interpretation. A CAD model on its own is a 3D shape, but the shape is designed in the light of knowledge about what the part must do, about how it can be certified as airworthy, and about how it can be manufactured. Retaining that knowledge is not as simple as taking a snapshot of the context, because it also depends on the knowledge of the user.

FIGURE 1.3 Plans for a single valve amplifier.

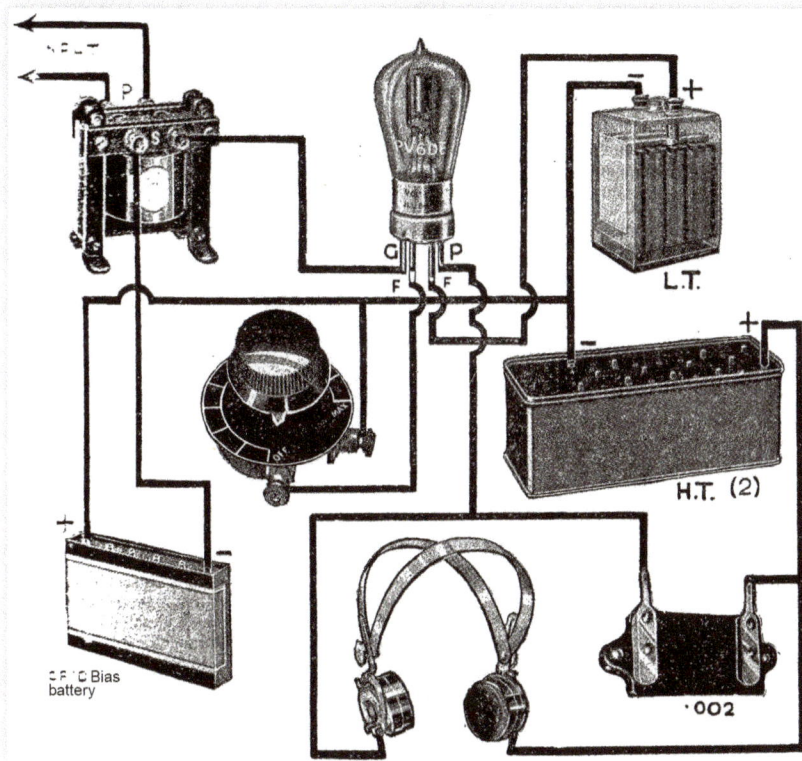

Consider for example, the diagram for a valve radio (Figure 1.3) from the 1928 *Every Boy's Hobby Annual* [1.4], a book expecting minimal knowledge from its readers: what is the component in the bottom right labelled ".002"? The diagram does not correspond to modern symbology. The text reveals it to be a condenser, an obsolete term for a capacitor, in units of "mfd," here meaning microfarad (now written μF) rather than millifarad (mF). Over the long term, knowledge retention is a moving target.

Established methods of media migration mean that data retention can be viewed as a solved problem. The main focus of the book is information retention—being able to reproduce usable facts about the design of an aircraft. Information retention focuses on checking that the software used to access old models reads those models correctly. Information retention also requires some level of knowledge retention since the context provides meaning to the information. This may be as simple as keeping a copy of the rules for numbering aircraft parts, or as complex as tracing through the cabinets of design standards for a process last used 50 years ago. Knowledge retention is a problem unique to each business, and all this book can do is give hints about what may be important.

1.4 Is It Long-Term Sustainment; or Retention or Archiving?

Sustainment? Archiving? Retention? Curation?—each term comes with hidden expectations, which make it easy to get into pedantic arguments about what to call the topic.

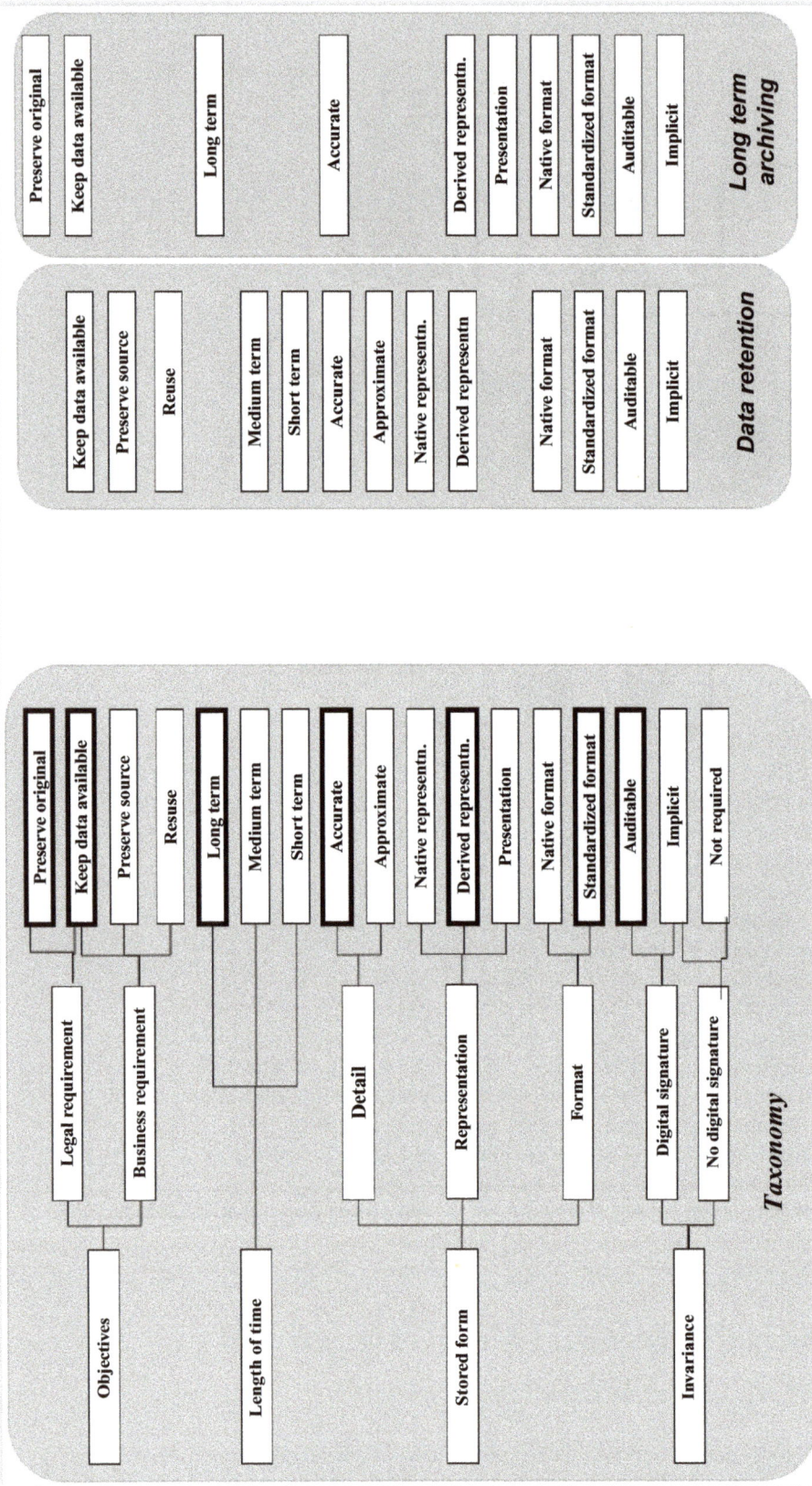

FIGURE 1.4 Concepts in long-term retention.

The underlying concepts are set out as a taxonomy in Figure 1.4, and typical choices are shown for *long-term archiving* and *retention* (based on [LOTAR 3][1]).

Archive too often suggests a dusty, forgotten room. Sustainment might be better as a reminder that the design information must remain available for reuse and reassessment. Terminology was an early discussion in the LOTAR project (see below and Chapter 5), and so it explicitly set out its scope as shown in bold boxes in Figure 1.4. In general, this book follows the LOTAR scope and will follow its usage to prefer "Retention" for the overall business process, although "Archive" is an easier usage for the system that stores the data and for the immediate interactions with that system, as in "long-term retention involves processes that cross the boundary between the business and the archive."

The business processes needed for information retention sit above the usual computer room processes of backup and archive, which act only on data and do not meet the criteria for long-term information retention. Cloud computing effectively virtualizes computer room operations and so is also a service for data retention rather than an information retention. Moreover, there is little guarantee that any particular Cloud service stack will remain usable over the longer term, as the various components in the stack each have their own, relatively short lifecycle.

Model-based design is contrasted with document-based design, where, rather than software, a human knows how to read the document—say, a drawing or a report—and manually interpolates the details required. Long-term preservation of documents has been extensively investigated in the libraries and the archiving sectors (see also Section 9.9). This sector often focuses on documents of record, such as government e-mails. In the academic literature, this is often referred to as data curation and has a strong emphasis on selecting the material to be preserved and on recording its provenance and context—who created it? when? who is allowed to read it?

Techniques for retaining design models have been developed by the LOTAR project, which is producing a standard for the aerospace sector. LOTAR is a collaboration based on the major US and European aerospace companies. It takes the exchange of mechanical CAD as the paradigm for long-term retention: there is a long history of exchanging CAD across space (between different sites), and long-term retention translates to exchanging data across time and what extra you need to do since you cannot ring 2019 from 2069.

Beyond the techniques, there are business and legal issues. For example, over time, companies merge or sell off parts of the business, and the company needs to ensure it retains intellectual property rights (IPR) of any data it holds. Some of these issues are touched on in the book.

1.5 **Sources for How: OAIS, LOTAR, RASSC**

One of the most influential documents on long-term retention is OAIS (the Open Archival Information System)—described in Chapter 3—which was developed by NASA to provide a reference model in order to compare different archiving solutions. Because it is a thorough and detailed analysis, it has frequently served as a template for developing archival systems. For example, the LOTAR project has based its detailed requirements on OAIS.

---

[1] [LOTAR 3] ASD-STAN, "Aerospace series—LOTAR—LOng Term Archiving and Retrieval of digital technical product documentation such as 3D, CAD and PDM data—Part 003: Fundamentals and Concepts," prEN 9300-003:2007 (also as NAS 9300-003 with the same title). As the LOTAR documents are referenced extensively throughout the book, the convention used is that [LOTAR 3] is part 3 of the EN 9300/AS 9300 standard.

LOTAR (LOng Term Archiving and Retrieval) is a project to standardize the retention of design models in the aerospace industry, starting with mechanical CAD. The origins of the LOTAR standard lie in separate US and European working groups dating from the mid-1990s, which later merged to form LOTAR International in 2003. Today the project includes participants from beyond the United States and Europe, and is tracked by other industries, including the nuclear and automotive sectors. The project and the international standard are described in detail in Chapter 5.

RASSC was a short UK project started some 10 years after OAIS was published, and looked primarily at economic and business models for long-term retention as a service (on the model of Cloud services). It has two features of interest here. First, it refactored OAIS into a service stack to understand how archiving-as-a-service might operate. In doing so it shows the different services needed for the retention of data, information, and knowledge. Second, it went beyond the current scope of LOTAR to investigate long-term retention to aircraft in-service data, and so links across to the world of Big Data.

These three projects represent the tip of a vast iceberg of deeply dull reports into the subject. They are used here to develop the ideas on long-term retention.

1.6 Summary

In model-based design, the file that stores the model is only a partial representation of the model—the other part is in the software that builds the complete model from the data. Consequently, the usual IT approach to archiving—save the bits and bytes—is inadequate, since the lifecycle of modelling software is shorter—sometimes much shorter—than that of the product being designed.

The ability to access original design information is important for three reasons. First, many products need certification to show that they are safe to use: in aerospace, this is the certificate of airworthiness. Aerospace certification authorities require ongoing access to design models else the certificate may be withdrawn. Second, design models may be required in a court of law, for example, following a crash, to show that the product was properly designed. Third, the models may be required for new design work: for modifications, repairs, or for reuse in the next product. Additionally, data from the in-service phase can be mined to improve the design, the support solution, and flight operation.

Consequently, over the last 20 or so years, substantial efforts have been made to develop ways of retaining models. In aerospace, this has been the work of the LOTAR project, which has taken forward the OAIS model of a digital repository, and has built on additional guidance for digital models. This includes techniques to demonstrate the integrity of the model—to show, not only has the data not changed, but more importantly, that the software has reconstructed the full model correctly.

References

1.1 Wikipedia, "Sarbanes-Oxley Act," accessed February 22, 2019, https://en.wikipedia.org/wiki/Sarbanes%E2%80%93Oxley_Act.

1.2 Barker, S., "LOTAR: High Integrity CAD Data for Aerospace," presented at *the Digital Preservation Coalition*, July 2010.

1.3 International Project Leadership Academy, "Airbus—A380," accessed February 3, 2019, http://calleam.com/WTPF/?p=4700.

1.4 *Every Boy's Hobby Annual 1928* (London: Amalgamated Press, 1928).

Why Retain Information?

2.1 Aerospace Business Drivers

On September 2, 2006, a Hawker Siddeley Nimrod suffered a fuel leak and consequently caught fire and crashed in Kandahar, Afghanistan, killing all 14 crew members [2.1, 2.2]. The aircraft had entered operational service in October 1969, and its airframe was based on the De Havilland Comet, an aircraft which first flew in July 1949 [2.3]. BAE Systems—the successor company to Hawker Siddeley—were directly blamed for failures in airworthiness, although the Ministry of Defence—the UK air-safety authority for military aircraft—were also criticized for prioritizing budget over airworthiness, and eventually admitted responsibility for the deaths following a legal action under human rights law. This example brings into play the primary requirements for long-term retention: airworthiness, legal liability, and reuse of design.

Were we to go into more detail, the example would point to secondary requirements such as the need to maintain in-service data—previously there had been fuel leak on another aircraft. Data from the in-service phase includes the manufacturing and service records, which witness the continuing airworthiness of the aircraft, and also health monitoring data, which records how the aircraft is functioning and warns of impending failure. There is only a limited need for the long-term retention of in-service data for analysis, but maintaining the interfaces to an in-service data repository has many features in common with archiving.

The example also points to the need to maintain data access in the face of industrial restructuring. For example, until 2000, the UK aerospace industry was consolidating into BAE Systems [2.4]. Since then BAE Systems has refocused, divesting itself of its Airbus division while becoming by acquisition a major US aerospace and defense contractor. An aircraft design starts out as an asset but could become a liability if the company that owns it is sold. Demonstrating that the design archive remains accessible could become a factor in industrial restructuring decisions.

This chapter reviews these business drivers and considers how they pan out as functional requirements for an archive.

2.1.1 Driver 1: Airworthiness

In order to fly, an aircraft must be certified as airworthy, and to start with, that means that the design of the aircraft type must be certified. In most countries this certification is provided by their own, independent statutory aviation authorities. In the United States, it is the Federal Aviation Administration (FAA), while across Europe it is the European Aviation Safety Agency (EASA).

For certification, an aircraft type covers those aircraft of a similar design [2.4]. Major manufacturers typically make a number of aircraft families, such as the Boeing 777 or the Airbus A330. Within each family they make several different models, for example, 777-200, 777-300. In practice, many aircraft models within the same family have a separate type code, although models which are sufficiently similar will be covered by a single code. A *type certificate* is issued by a regulatory agency (FAA, EASA, etc.) informed by the *type design*, the operating limitations, and any other limitations or conditions [2.5].

According to the code of Federal Regulations, type design [2.5] consists of:

a. The drawings and specifications, and a listing of those drawings and specifications, necessary to define the configuration and the design features

b. Information on dimensions, materials, and processes necessary to define the structural strength

c. The airworthiness limitation section of the Instructions for Continued Airworthiness

d. Conditions applicable to pilot-owner

e. Any other data necessary to allow, by comparison, the determination of the airworthiness of later products of the same type

In the United States, the FAA maintains a *custodial record* of the type design [2.6]. To submit data electronically [2.7], the aircraft manufacturer should obtain a memorandum of agreement with the FAA, which must answer a number of issues, the first of which is "***How will the data integrity be assured throughout its life cycle?*** *Include a process to verify and ensure that data integrity and continued access is maintained during hardware and software upgrades and conversions, system maintenance, and other life-cycle activity. … Type design data must be retained and accessible for the lifespan of the product. It is possible that technical support for the original software will be terminated during the product lifespan, so your procedures manual must explain how access to the data will be retained or transitioned to a new software system*" [2.7]. This is followed by a number of other questions including:

- How will users access data at its point of use? The process should describe the availability to remote users such as the FAA.

- How will the configuration of the final product be established?

- How will the electronic type design data be transferred to the FAA when the holder ceases to operate?

- How will the FAA and other regulatory agencies get access to the data?

- How will users be trained in the 3D modeling system? Describe for all users including the FAA.

Essentially, the FAA circulars appear to be taking the very reasonable position that if an aircraft manufacturer wants to keep electronic design records, then it is up to them to sort out data sustainment. The FAA wants to be able to read the data but not to pay for the upkeep of the archive or migration to new software. The FAA requires that the "engineering intent" is preserved for the life of the product flying under that type certificate, and it is up to the product approval holder (PAH) to document a process to maintain that information over the lifecycle of the data.

2.1.2 Driver 2: Product Liability

Product liability is a legal minefield. In the United States, an aircraft manufacturer has strict liability for a design or manufacturing defect, meaning that it is the defect rather than negligence which leads to liability. However, there are concerns that some state laws can override federal law, so that conformance to FAA guidelines may not be enough to prove lack of liability [2.8, 2.9]. In the United Kingdom and some EU states, strict liability also applies based on their implementation of the Product Liability Directive 85/374/EEC [2.10]. Liability is further complicated by long, transnational supply chains, which may make it more lucrative to sue in one jurisdiction rather than another.

From the data retention viewpoint, this creates at minimum a requirement to retain design information. A business may also require active curation of the data, so that when data is no longer legally required it is deleted, since then that data cannot be interpreted by a plaintiff as showing negligence.

However, the ease with which electronic data can be altered creates additional requirements if it is to be used as evidence. At minimum, this is likely to include evidence that it has not been tampered with in storage. In the United Kingdom, print-outs of electronic documents are classed as "hearsay," and may not be admissible unless provided with "evidential weight" that they are what they purport to be [2.11]. Electronic documents can easily be thrown out of court when the so-called compliance points are found to be missing in systems from which a company sources evidence. These include:

- No information policy document

- No retention schedule

- Inappropriate security controls

- Lack of procedural documentation

- Insufficient control on document input procedures

- Insufficient information about the technology from the system supplier

- Use of inappropriate facilities, such as image cleanup

- No thought of future migration requirements

- Lack of documentation on audit trail content and access procedures

For example, the fact that a document has been held unchanged in an archive does not prove that it was the output of design process—the process definitions need to follow through from design approval to submission to the archive. The United States has a similar definition of "hearsay" to the United Kingdom, and both countries' legal systems are based on common law, whereas much of continental Europe have systems based on statute law, meaning that courts have no power to act unless there is an applicable statute.

Given that aircraft operate through many jurisdictions, this requires that an archival system is constructed to meet legal obligations such as evidential weight.

2.1.3 Driver 3: Reuse

A priori, one would expect that reusing an existing design model should be cheaper than regenerating it from scratch. However, between one generation of a modeling system and the next, major changes within the software will create uncertainty as to whether a migrated model is identical to the original. The generation change mentioned in Chapter 1 illustrates this.

In aerospace, the quality controls and airworthiness checks constitute a major element of the cost of a design change. Migration to a new generation of CAD may require these quality checks to be redone on each model, and manufacturers may be reluctant to bear the cost of migrating all current models. They may opt instead to avoid migration unless the part modeled is to be modified. The business case can vary between aerospace sectors. For example, in the military aircraft business, a new aircraft is likely to have a completely new airframe, possibly using new materials, so there is little reuse of design from one type to another. In such an environment, the cost of remodeling may be included in the cost of the change (and therefore chargeable to the customer). The case for reuse is likely to be stronger in the commercial sector, where a series of models in a design family may be developed over several years.

2.2 Retaining Data across the SupplyChain

Aircraft manufacture involves long supply chains, with assemblies built up from components obtained from suppliers, and which in turn are assemblies of components from their suppliers, and so on. This in turn creates a supply chain of information—shapes of components, weights, materials, electrical paths, and so on. It also creates a chain of responsibilities for airworthiness and therefore of liability—even simple fixings such as rivets may require to be certified.

This leads to two conflicting requirements: that supply-chain information is accessible and that proprietary information is kept confidential. From the viewpoint of the aircraft manufacturer, they will want assurance that the suppliers' information will remain available and accessible for the life of their aircraft. Here there are three options:

- The manufacturer obtains and retains all the relevant supplier information
- The supplier builds its own trustworthy archive
- The supplier uses a trusted third-party repository

The supplier will be concerned that the first option may compromise their business if, in future designs, the aircraft manufacturer tenders to rival suppliers. The second may impose substantial costs on the supplier, and the aircraft manufacturer will be concerned to ensure that the suppliers' archive meets long-term retention standards and how access would be maintained should the supplier cease trading or be taken over. The third option would deal with most of these concerns if a suitable data retention service company could be found—the implications of this option is considered further in Chapter 4.

2.3 Retaining Data through the In-Service Phase

The in-service phase uses manuals and procedures from the design phase, collects vehicle health monitoring data, and applies maintenance procedures to monitor the aircraft, keep it airworthy, repair it, and apply updates and modifications. Some modifications can be extensive, such as converting a passenger jet into a cargo plane. Manuals and procedures may be provided as electronic documents and which may include 3D visualizations. Such 3D visualizations are typically derived from CAD models, so reissue of documents—which may follow from new viewing technology—provides a further reason to retain and reuse CAD models.

Most health monitoring data and maintenance records have a fairly short life but there are some where long-term accumulation of the data may be required. Here the requirement will be less focused on sustaining data than on sustaining data services. For example, structural life assessment for composite wings uses sensors embedded within the structure of the wing, and data from each flight is downloaded after each flight [2.12]. The requirements here are both that the data is accumulated over the life of the aircraft and that the services that download the data to the archive last for the life of the aircraft—one cannot update the sensors without replacing the wing. Health monitoring will include systems designed by suppliers, and concerns about access and confidentiality will be similar to those described in Section 2.2.

2.4 Retention and Business Restructuring

The US and European aerospace industries have seen extensive restructurings over the last 30 years, and organizational responsibilities for product data may be transferred several times over the life of a product. From the long-term retention viewpoint, this implies the information once held in the archive of one company must be held in the archive of a different company. This may simply mean re-badging the existing archive with the new company's logo, or it may require splitting an existing archive in two, or transferring data to a new archive in the new company.

From a business perspective, the prospect of restructuring will involve confirming the existing archives meet the long-term retention requirements for airworthiness, legal liability, and reuse, otherwise the new company may have to take on the costs of bringing the archive up to standard. After the restructuring, the new company may need to deal with multiple archives. Here, conformance to a common standard will make management of the archives simpler and less expensive.

2.5 Quality Requirement: Key Characteristics

The preceding sections concentrated on business drivers, particularly why design data needs to be retained well into the future. But what quality criteria that apply to data retention? For example, the yellowing of paper over time does not lose the contents of a document, but the loss of pages does, so a quality criterion for a physical archive will be to ensure all pages are retained. For electronic data, the criterion is the preservation *of key characteristics.*

A *key characteristic* is a feature of a material or part whose variation has a significant influence on product fit, performance, service life, or manufacturability [2.13]. For example, if a part in a mechanism was manufactured to larger tolerances than specified, although the part is the right shape, the mechanism might jam—that is, *fit* is a key characteristic, and any record of the part must include details of manufacturing tolerances. Note the inclusion of service life in the definition, since if a part has a shorter service life than the certified design, then it will need a shorter inspection/replacement period. The essential quality criterion for data retention is therefore that the models and descriptions of key characteristics are retained unchanged. Identification, measurement, and recording of key characteristics is addressed in the LOTAR standard.

2.6 **LOTAR Requirements**

Starting from the business drivers of airworthiness, legal liability, and reuse, the LOTAR project has set out the overall functional requirements for an aerospace sector archive [LOTAR 2].

Under the heading "Acceptance" [of the archive] it requires that a document must be retrievable for at least a planned retention period, that it can be accessed within a set time, that the data is available for further use and manipulation (perhaps with reduced functionality), and that the lifecycle costs for archiving be minimized.

Legal requirements include the ability to show that a product conforms to its archived documentation—essentially that all the relevant documents are archived—that it meets the relevant laws on data security and privacy, and that the system be auditable. The requirement on privacy may seem odd at first; however, many aerospace products must be authorized by a suitably qualified person. This, in turn, requires that the person is identified along with evidence of their qualification, and perhaps even evidence that they were employed in that role at the time of authorization. This information can raise privacy concerns under some jurisdictions.

There is an extensive list of requirements under "Security and Integrity," which can be summarized as follows:

- It shall be impossible to destroy a document over its scheduled lifetime.

- Each document will be protected against unauthorized changes and the information content remains unchanged.

- The archive will provide proof that the document is unchanged.

- The archive will allow (where feasible) checks on key characteristics.

- Audit trails will provide evidence that archiving processes have been followed.

- The archive will allow migration to new platforms, media, and software.

- The risk and expense of running the archive will be managed.

- The archive will manage access rights including providing audit trails.

The standard then goes into further detail, based on the OAIS model (see Chapter 3). Although the OAIS model is a descriptive model designed to compare archives, LOTAR reexpresses the model as normative requirements—essentially saying, take the OAIS model and implement it.

2.7 **Summary**

The main business drivers for long-term retention are airworthiness to get the aircraft into the air, legal liability once it is flying, and design reuse to start the next variant or aircraft. Much of the focus on retention is on design data in support of obtaining a type certificate of airworthiness.

Secondary drivers include access to data along the supply chain so that the full set of design data is maintained; retention of in-service data requiring that the services that process the data are sustained; and business restructuring, which requires the transfer of archive data between companies. These three drivers point in the direction of a sustainable "archiving-as-a-service" provided by a third party (Chapter 4).

Retention of *key characteristics* is the main quality requirement. LOTAR provides a detailed breakdown of these requirements.

References

2.1 Wikipedia "2006 Royal Air Force Nimrod Crash," accessed January 25, 2019, https://en.wikipedia.org/wiki/2006_Royal_Air_Force_Nimrod_crash.

2.2 Haddon-Cave, C., *The Nimrod Review: An Independent Review into the Broader Issues Surrounding the Loss of the RAF Nimrod MR2 Aircraft XV230 in Afghanistan in 2006* (London: Stationery Office Books, October 28, 2009), ISBN: 9780102962659 (Aka "Haddon-Cave Report"), 587 pages found at https://assets.publishing.service.gov.uk/government/uploads/system/uploads/attachment_data/file/229037/1025.pdf, accessed January 25, 2019.

2.3 Wikipedia, "de Havilland Comet," accessed January 25, 2019, https://en.wikipedia.org/wiki/De_Havilland_Comet.

2.4 FAA-Aircraft-Certification.com, "'FAA Definitions' (Type-2)," accessed January 28, 2019, http://www.faa-aircraft-certification.com/faa-definitions.html.

2.5 SAE "A Process Standard for the Storage, Retrieval and Use of Three-Dimensional Type Design Data Sets," SAE AS9034, Draft 2003-03-03, 8.

2.6 FAA, "Certification Data Retention Agreements and Government Records," FAA Advisory Circular 20-179, June 10, 2013, accessed January 28, 2019, https://www.faa.gov/documentLibrary/media/Advisory_Circular/AC_20-179.pdf.

2.7 FAA, "Using Electronic Modeling Systems as Primary Type Design Data," FAA Advisory Circular 21-48, October 29, 2010, accessed January 28, 2019, https://www.faa.gov/documentLibrary/media/Advisory_Circular/AC_21-48.pdf.

2.8 Abbey, D., "Aviation Product Liability—Are You at Risk?," accessed January 29, 2018, https://integrogroup.com/uploads/white_papers/2015-04_Aviation-Products-White-Paper.pdf (Abbey works for Integro, a company specializing in aviation risk and insurance).

2.9 Epstein, C., "Courts Say FAA Not the Final Word in Product Liability," May 16, 2017, accessed January 29, 2019, https://www.ainonline.com/aviation-news/general-aviation/2017-05-16/courts-say-faa-not-final-word-product-liability.

2.10 Ashfords, "Laws and Litigation: Aviation," accessed January 9, 2019, https://www.ashfords.co.uk/media/3001/aviation-booklet-laws-and-litigation.pdf.

2.11 BS 10008:2014, "Evidential Weight and Legal Admissibility of Electronic Information."

2.12 Barker, S., "The RASSC Project," in Jennions, I.K. (ed.), *Integrated Vehicle Health Management: Implementation and Lessons Learned* (Warrendale, PA: SAE International, 2015), ISBN: 978-0-7680-8088-9.

2.13 EN 9100:2003, "Quality Management Systems. Requirements for Aviation, Space and Defense Organizations," AS 9100 which in turn is based on ISO 9001:1994-As cited in LOTAR's "Part 007: Terms and References," EN 9300-007:2013. Note EN 9100:2018 supersedes the 2003 edition.

3

OAIS: The Model for an Archive

3.1 What Is OAIS? A User Perspective

The OAIS report describes "*a common framework of terms and concepts which make up an Open Archival Information System (OAIS). It allows existing and future archives to be more meaningfully compared and contrasted*" [3.1].[1] It was created for NASA by the Consolidated Consortium of Space Data Systems (CCSDS) and was adopted as ISO 14721 in March 2003. In practice, for a new user, it provides a comprehensive guide to all the things to look for in an archive, and for an experienced user, it gives a way of assessing what any particular archive provides. And, as it is a thorough analysis, it has also been used as a blueprint for an archive.

The OAIS concept of an archive is of an organization, comprising people and systems, with responsibility to preserve information and make it available to a *designated community* [3.1, p. 11]. As well as physically preserving information, it is responsible for answering two key questions:

- What information must the archive collect and store to meet the needs of the designated community?

- How must the archive respond to changes in technology and in user environment to continue to meet those needs?

For example, the aerospace requirements envisage three separate designated communities for aircraft CAD models:

---

[1] The 2012 edition of [3.1] is a recommended practice, whereas the original 2002 edition was a recommended standard. The CCSDS is an association of space agencies, including NASA.

- The aviation authorities, who need to certify the aircraft

- Lawyers and their experts, in the investigation of product liability cases

- Company design engineers, who may reuse some elements of a design

The first two communities will need only to review the design from time to time. For them, the archive might store a 3D visualization of each CAD model and would maintain a visualization portal so that the aviation authorities or lawyers do not need to buy visualization software. As older visualization software becomes obsolete, the archive would have the responsibility of replacing it and providing training for the new software, so meeting some of the demands of the FAA (cf. Section 2.1.1).

In the case of design engineers, they would need access to the full models to reuse them but would have their own CAD software and training. Here the archive would need to track file formats used by the designers, and may need to migrate stored CAD models from old formats or to provide a conversion facility when a model is requested.

Working backwards, the original designers would provide the archive with a package containing both a CAD model and its visualization. They would also need to identify the CAD file format should conversion be needed at a later date, and also provide the archive with a set of search terms that would enable users to find the files they need.

Thus, the archive has roles in both corporate memory and as a technology consultant. The latter role includes watching for changes in CAD technology and finding routes through from old technologies to newer ones.

Of course, the devil is in the detail, and OAIS has clearly led to many working like demons to fill in that detail. This is found not only in the OAIS specification, but also in the guidance on what information the creators need to submit [3.2], the metadata specification [3.3], a checklist to assess if a repository is trustworthy [3.4], and so on [3.5].

This chapter briefly summarizes OAIS from the perspective of aerospace companies. It is assumed that "The Archive" is a function within an aerospace company, as opposed to an independent body. This means some detailed issues will not arise—for example, a major challenge for an independent archive is obtaining the rights to use the information, but that does not arise where the industry is archiving its own information.

3.2 **Key Processes in OAIS**

Figure 3.1 shows an overview model of how the archive works. Boxes with a white background are taken from the OAIS Functional Model ([3.1], Figure 4.1); the other boxes have been added for clarity. The model here follows the convention that products are produced by processes that are staffed by an organization. The fourth element of that convention—project—is not applicable here as the processes are activated many times over the life of the archive.

The elements of the model are:

- Organizations, shown by the swim lanes (background rounded rectangles):

  - The producer, who creates the information archived

  - The archive, which stores and indexes the information

  - The consumer, a user in one of the designated communities who accesses information

FIGURE 3.1 A model of an archive.

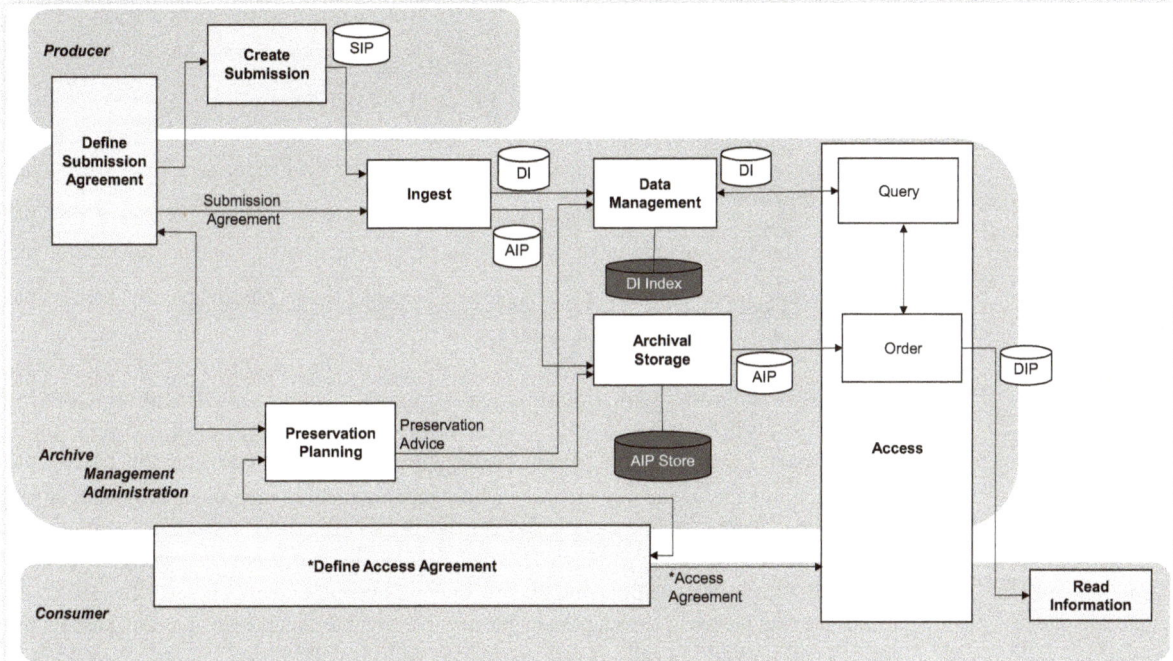

- Products are annotated against the flow lines between processes, including the OAIS *information packages* (shown as cylinders):

  - Submission Information Package (SIP)—provides the content information for archiving

  - Archival Information Package (AIP)—what is actually stored

  - Dissemination Information Package (DIP)—the information the consumer receives

  - Descriptive Information (DI)—describes what the packages contain

  - Submission Agreement—describing the SIP and the producer/archive interface

  - Preservation Advice—to sustain the information for future access

  - *Access Agreement—describing the DIP and its conditions of use

- Processes (boxes)

  - Define Submission Agreement—identify what the archive needs from the producer

  - Create Submission—the producer packages their information as a SIP

  - Ingest—take in the SIP, add metadata to create an AIP, and extract the associated DI

  - Archival Storage—store the AIP and refresh media as technology changes

  - *Define Access Agreement—identify what the consumer gets from the archive

  - Access, with two subprocesses

    - Query, to find information

    - Order, to access information and repackage AIPs into a DIP

- Data Management—storing the DI and answering queries against it
- Preservation Planning—providing advice on sustaining the information

Items prefixed * are extensions to the OAIS model.

Figure 3.1 illustrates three main stories: how information gets into the archive (Ingest), how consumers get to read that information (Access), and how the archive maintains its usefulness in the face of change (Preservation Planning). These are further expanded below.

3.2.1 Ingest—How Information Gets into the Archive

Before any data is ingested, a Submission Agreement is created, identifying what the producer wants to be archived and the needs of the designated communities. For example, "one million CAD files will be submitted over a five-year period to be available to Aviation Authorities for the life of the aircraft." This agreement must also meet the needs of the Archive on data format, for example, "a CAD file in ISO 10303-203 format (to minimise the effects of file format changes) plus a 3D visualisation file." It must also cover the DI needed to identify the file, for example, the part number and version, the NATO stock number (for a military aircraft), and the change order number (which is important in confirming the configuration of an aircraft). It also covers the technical aspects of the interface, such as how the archive authenticates that the sender of the information is actually the producer.

For each ingest, the producer collects the target data they want to archive and packages this together with DI into a SIP. On Ingest, the archive does agreed quality checks, for example, that the SIP does actually contain both a CAD file and a 3D visualization file. These could be submitted in separate SIPs, but then the archive would have to deal with the situation that it has sent one file but is left waiting for the other, but cannot archive either until both are received.

The Ingest process then creates AIP and the DI for each package. In the case of a CAD archive, the CAD file and the 3D visualization are likely to be split into separate AIPs because the different designated communities want one or the other but not usually both, and the different formats may have different lifecycles and therefore format migration will occur at different times.

The AIPs will also have representation information added to enable the file to be decoded in future. For example, the ISO 10303-203 format includes an identifier of the format within the file itself, so that does not need to be added, but the AIP will also need a link to the ISO 10303-203 standard (at 500 pages, it would not be practical to include the standard in a million AIPs). This linkage will be defined in the submission agreement.

Each AIP also has preservation description information added—this is described further in Section 3.4. The AIPs are then sent to archival storage. Drilling down into archival storage, there are processes such as media refresh, which moves the files to a new physical store as current technology becomes obsolete (see also Chapter 5).

The DI—also called a package description—contains information the consumer needs to find and order the content. This is an area where OAIS is necessarily rather vague. What information consumers use to find a particular CAD model and how they order a package will ultimately come from the various designated communities. Since a consistent architecture will show the products that record this information, the OAIS model is extended here to include an *access agreement, which is described in the next subsection. The DI is stored and indexed in the data management process.

3.2.2 Access—Getting It Out Again

Although OAIS defines an archive in relation to its designated community, it has to take a very general view of archiving in which the designated community is ill defined. An aerospace archive will serve three well-defined designated communities: Design, Aviation Authorities, and Legal. These communities come with clear requirements, and it therefore makes sense to formalize these requirements in an *access agreement stating what information they need, its form, and the access interface. Setting out the *access agreements is therefore the first step to providing access to the archive. Other communities might be added at a later date.

The second step, the access process, has two subprocesses. The first subprocess query allows the consumer to search the data management indices to find the information that is of interest. The results of this process are fed through to the subprocess order, in which the consumer asks for particular information packages, and which the archive provides as a DIP. A key requirement here is that the information provided is understandable by the consumer. In the first instance, this means either delivering the information in a form the consumer's tools read or providing a remote access service. By analogy with a physical archive, a library which stored documents on microfiche would also provide microfiche readers on its premises.

Access should also come with a second level of service, delivering the supporting information needed to understand the content information. If what is stored is a manufacturing drawing, then the supporting information would cover, for example, the symbology and conventions for drawing screw threads or specifying manufacturing tolerances (e.g., "Readers of CAD 123 also ordered *ASME Y14.41-2003 Digital Product Data Definition Practices*").

3.2.3 Preservation Planning—Keeping It Live Inside the Archive

All models fall foul of the limited set of abstractions that any particular modeling technique allows and preservation planning exposes the limitations of the process/product/organization model of Figure 3.1. Preservation planning is better viewed as the combination of a body of knowledge and a set of processes to refresh or apply that knowledge. These processes are event driven, they run in parallel, they feed back into each other, and they run over different timescales. The clearest way to treat preservation planning is to look at what it does in the long, medium, and short term.

The long-term goal lies in preserving existing collections, planning for new collections, controlling the cost of the archive, and controlling the risks of technology obsolescence. Part of that cost is training and retaining people knowledgeable about data sustainment. For example, understanding a complex information standard such as ISO 10303-203 requires several months of technical work. The outputs from this knowledge are operating policies for the archive, preferred information standards (to minimize the number of formats that archivists must understand), and reviews of how the archive is operating.

As part of its medium-term strategy, the archive needs to monitor changes in technology—not only the technologies used for storage within the archive, but also the technologies used by the producers and understood by the designated communities. Although to the casual observer, technology change may seem rapid, in practice the period from the development of a new technology to its widespread adoption is a matter of several years. In aerospace, such monitoring feeds back into both the acquisition of new knowledge and the evaluation and purchase of new technology. It will also feed into long-term policy development and may also feed across to the producers and consumers by way of advice on what will be sustainable for the life of the aircraft, and therefore into the revision of the submission and *access agreements.

The short term is occupied with the development and implementation of new submission and access agreements. These will take into account preferred technologies and standards in order to contain the costs and risks to the archive, but will also feed back user requirements into the medium- and long-term thinking. These agreements will also form part of the *representation network* (Section 3.4) needed to sustain the archive and the information it contains.

It is likely that there will be conflicts between what the users see as the ideal technology for their working environment and what the Archive sees as sustainable over the life of the aircraft. Resolving such conflicts is a matter for corporate governance (Chapter 6).

3.3 Metadata—Remembering What the Archive Contains

Metadata—literally data about data—provides a record of the things you need to know in order to use your target information effectively. For example, in order to use a CAD model, at minimum you need to know what part it refers to, whether it is authorized for use, and what CAD package can access it. However, metadata is a very broad term, and the way OAIS interprets it is not always obvious from a user perspective. The approach here is to start by differentiating three types of archival metadata: technical, business process, and supporting information.

Technical metadata is used by software to decode the content of a package or a file, rather than being used by a reader to interpret the contents. There are two subcategories of technical metadata. First, representation data informs software accessing the target data how to read it—for example, an image file may consist of a list of pixels along with the metadata of image size (W pixels by N pixels) used to break the list into N rows W pixels wide. Second, packaging information is used by the archival software to build and dismantle data packages—the SIP, AIP, and DIP. Figure 3.2 [based on 3.1, Figure 4-18] annotates the OAIS view of an information package to pick out the main classes of metadata.

Supporting information lists the supporting documents needed to make sense of the target data in the long term. This will include things such as details of file formats or drawing conventions—information that is part of the day-to-day knowledge of the people and software producing the content information, but which may be superseded and lost in the medium to long term. Supporting documentation is treated in the next section.

Users interact with data through four business processes. In the first two, users find target data and check that they can trust that it is the right data. The information management system runs the other two by managing the data and applying usage controls. These metadata needed by these processes can be summarized via the "MUST of metadata":

- Manage—e.g., identify the data, set review and expiry dates, link to change control orders

- Usage control—access control, intellectual property rights, and so on

- Search—alternate identifiers, descriptions, who created it, when, and so on

- Trust—approvals and authorizations, checks that the data is unaltered and so on

FIGURE 3.2 Information package data structure.

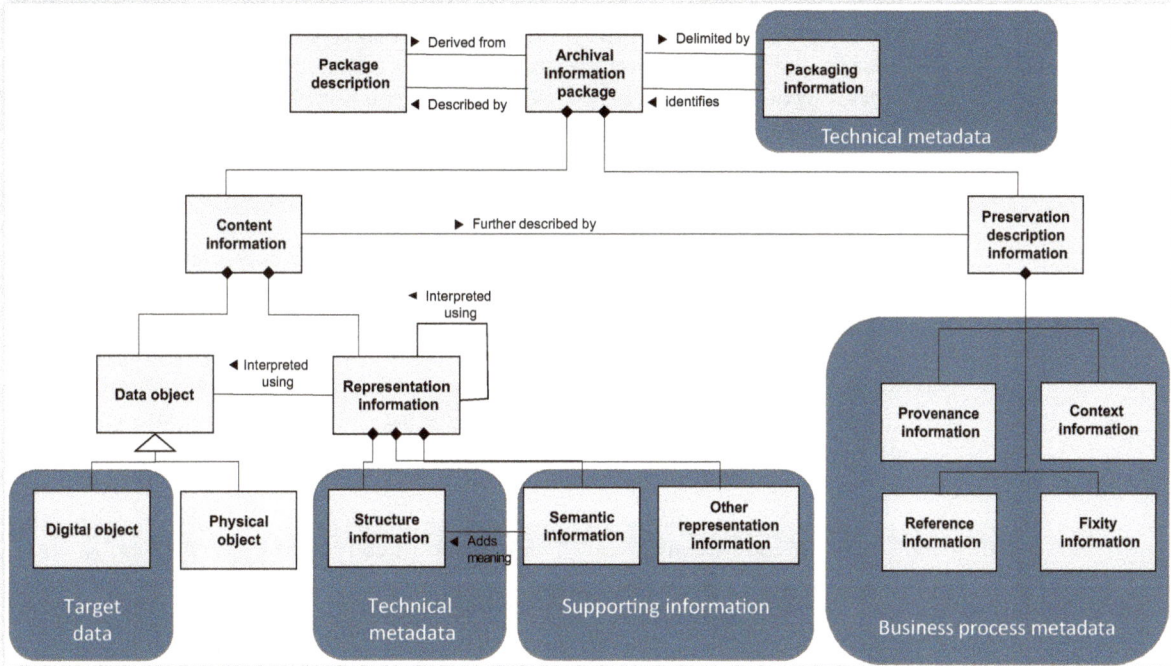

OAIS treats metadata from an archival perspective rather than the user perspective, but this disjunction could create a risk that the archive will miss user requirements. To smooth over this disjunction, Figure 3.3 maps the user "MUST" perspective to the classes given by OAIS.

A detailed list of archival metadata definitions has been developed by the PREMIS working group (Preservation Metadata: Implementation Strategies) [3.3, 283 pages],

FIGURE 3.3 Mapping information management metadata to OAIS categories.

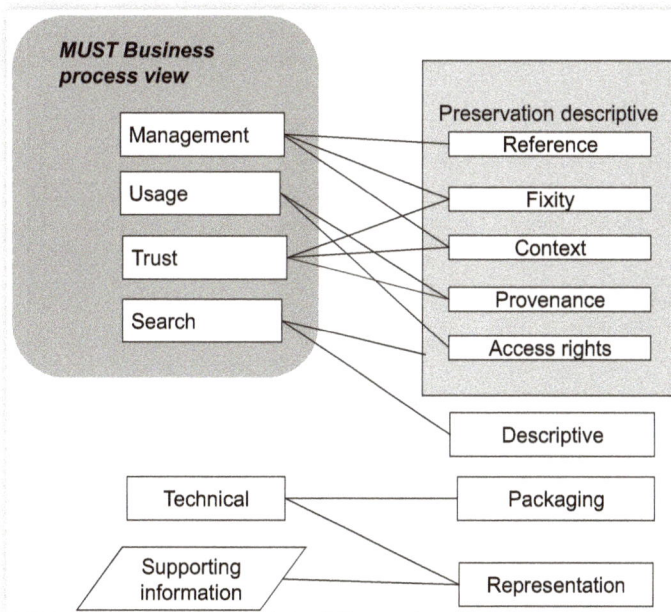

but this is focused on the archive's processes. Users, quite rightly, resent having to enter metadata that is vague or irrelevant to them. Where metadata is not useful to the producer but is required by the archive or by the designated communities, its entry must either be enforced by rigorous quality controls or it should be entered automatically or entered by dedicated data curation clerks. Failure to heed the user perspective risks the metadata being populated with meaningless or random values. The reverse mapping of that shown in Figure 3.3 can help to identify such risks.

To manage the design data for a million parts, an aerospace archive will need to provide metadata that maps to the industry's processes and software tools. For example, *part number* is used to identify a part, to link parts to assemblies, and as the anchor for all part-related documentation. In aerospace, *part number* is provided by an enterprise identifier prefixed by the organization that assigned it (Cage, DUNS, or ECC.UCC) and the part identifier for the configuration controlled design [3.6], for example, "CAG K6807 F100200300400AP." In aerospace, one candidate format for such metadata is the PLCS standard [3.7]. Focusing down onto what data is needed by aerospace and where it overlaps with PREMIS definitions would be a substantial study in its own right.

3.4 Metadata—Remembering What the Data Means

The question: if some time in the future you had to recover design information for a 25-year-old product from a division that closed down 10 years ago, what would you need to know? A first pass at answering that for a CAD file is shown in Figure 3.4 [based on 3.7].

The top left quadrant of Figure 3.4 shows the target data—a CAD file with embedded validation properties for some key characteristics. The CAD file embeds the serial number

FIGURE 3.4   Supporting information for a CAD file.

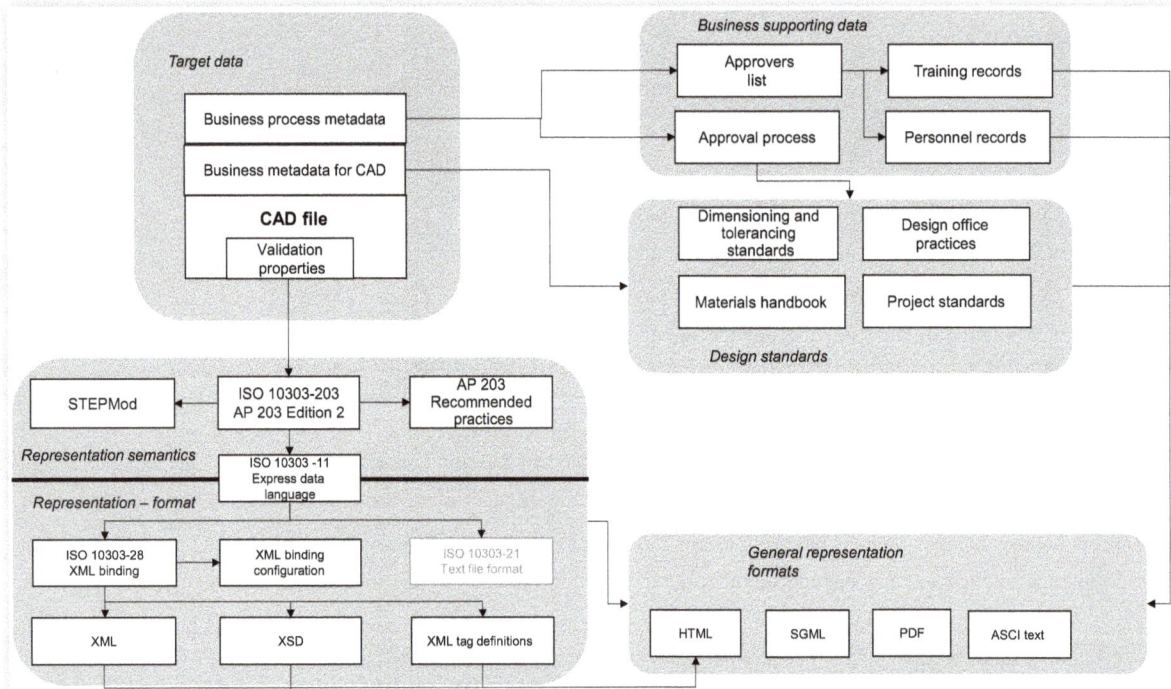

of the part, but this probably needs the enterprise identifier adding (see Section 3.3) to disambiguate it from parts created by other companies in the collaboration. The target data will also include metadata associated with the CAD model, notably identifying the CAD development process and the names of the people who checked and approved the file.

For the airworthiness community, links to the design process and design standards are needed to show the criteria for approval. The legal community will want the names of the approvers, along with proof they were suitably qualified and that they were actually working for the company at the time. This supporting data is shown in the top right quadrant. It is assumed that the archive is pre-populated with this data, and that the SIP identifies the data sources so that the required links can be created within the archive.

The lower half of the figure shows representation documentation—in OAIS terms, a combination of structure, semantic, and other representation information. The lower right quadrant shows some common document representation standards that may be assumed to be part of the archive. The lower left shows the representation documentation for the CAD file. This is divided into representation semantics and representation format. Representation semantics (above the dashed line) describes what the data means in the business, for example, identifying the entity *line* as a straight line specified by its end points. The representation format (below the dashed line) shows how the semantics are represented in a file—for example, a straight line might be written

```
#10 = point('P01', 2.0, 2.9);
#20 = point('P02', 5.0, 2.9);
#30 = line('L01', #10, #20); /*#10 and #20 refer to the
two end points*/
```

These supporting CAD standards are themselves archived using one of the general representation formats in the lower right. This illustrates that supporting information often expands into a *representation network* as one repeatedly asks the question "what do I need to know to understand this information?" And, of course, this representation network is written from the perspective of current knowledge, and it is one of the jobs of preservation planning to review this as the knowledge of the designated communities changes.

3.5 **One Archive or Many?**

OAIS section 6 on archive interoperability does not offer any normative guidance on terminology for archive interoperability, but rather reviews the technical and managerial issues that need to be resolved for four different interoperability types:

- Independent—no interoperability

- Cooperating—with common submission and dissemination standards, but without common access services

- Federated—with common access services

- Shared resources—based on a functional breakdown of the archive, rather than the service interface view of the previous three categories.

For aerospace, there are two main viewpoints on archive interoperability, that of a user in a designated community and that of an aircraft company operating its own archive.

For a member of a designated community, archive interoperability essentially means that archives have the same access interface and that similar types of model from different archives can be accessed via the same user interface. For example, where the aircraft company and its suppliers each have their own archives, an aviation authority could use a single log in to access any of the archives, it could search for a document across all of them without having to know which archive holds the document, and it could read any document through the same document viewer. This corresponds to *federated* interoperability. They might also like to see 3D models in the same space, so they can see how they fit together, but that is a requirement for *model interoperability*, which goes well beyond archive interoperability.

For aircraft companies operating multiple archives, interoperability requirements can include common submission and *access agreements, interoperable data management systems, and interoperability of internal mechanisms. The information packages and DI would likely be required to meet common standards. This would reduce the costs of an archive and would support a variety of business requirements such as the ability to merge or split archives following business restructuring, or for a company to hold records from a lower-tier supplier in the event of them ceasing to operate. This corresponds to the *cooperating* archives.

Although OAIS describes types of archive interoperability, it does not go so far as to mandate the detailed standards that would be required. For example, earlier sections of OAIS state that there is a submission information package and that it needs packaging information; however, OAIS is not an information standard for that packaging information or for the structure of the SIP.[2] Rather, under interoperability it notes only that where one archive ingests information from another, the producer archive would output a DIP that would be a SIP for the consumer archive, which implies both use a common packaging format. Ensuring interoperability across aerospace archives would need a separate, industry-specific standard. Note that LOTAR (Chapter 5) is producing industry-specific standards on content information but has not yet considered archive interoperability.

3.6 **Summary**

OAIS sets out what an archive looks like and provides a very useful set of common terminology. It also gives quite a lot of guidance on things that an archive implementer should think about, and there is a plethora of associated resources that get into the detail, such as what a submission agreement covers. And it also shows that an archive is complicated enough to need a significant investment in time and effort to develop a working knowledge of what archiving entails.

References

3.1 Consultative Committee for Space Data Systems (CCSDS), "Reference Model for an Open Archival Information System," Recommended Practice CCSDS 650.0-M-2, June 2012, http://www.ccsds.org/.

3.2 ISO 20652:2006 (CCSDS 651.0-B-1:2004), "Space Data and Information Transfer Systems—Producer-Archive Interface—Methodology Abstract Standard (PAIMAS)."

---

[2] Several candidate standards are cited in OAIS appendix D, and CCSDS also publishes a number of other related standards.

3.3 PREMIS Editorial Committee, "PREMIS Data Dictionary for Preservation Metadata," version 3.0, June 2015. Revised November 2015 superseding "The OCLC/RLG Working Group on Preservation Metadata," Preservation Metadata and the OAIS Information Model June 2002, http://www.oclc.org/research/pmwg/.

3.4 ISO 16363:2012 (CCSDS 652.0-M-1), "Space Data and Information Transfer Systems—Audit and Certification of Trustworthy Digital Repositories" and ISO 16919:2014 (CCSDS 652.1-M.2), "Space Data and Information Transfer Systems—Requirements for Bodies Providing Audit and Certification of Candidate Trustworthy Digital Repositories."

3.5 OAIS Reference Model (ISO 14721), "OAIS Roadmap and Related Standards," http://www.oais.info/standards-process/oais-roadmap-and-related-standards/, accessed February 7, 2019.

3.6 ISO 21849:2006, "Aircraft and Space—Industrial Data—Product Identification and Traceability."

3.7 ISO 10303-239, "Product Life Cycle Support."

3.8 Based on Barker S "Long Term Data Retention"— probably presented at PDT Europe c. 2010—but lacking provenance metadata.

Archiving as a Service

4.1 What Is *Archiving-as-a-Service*?

The previous chapter described the OAIS model of an archive. Here we look at one way of implementing an archive: *archive-as-a-service*. That is, rather than answering the question "How do we *build an archive*?" we look at "How do we access *archiving capabilities*?" Here, rather than follow a left-to-right process from producer to consumer, archiving is broken down top-to-bottom into services, with the simplest services at the bottom, and the most knowledge intensive at the top. This exposes much detail without needing to explain it at tedious length.

There is also a significant business reason for considering archive-as-a-service. Aerospace companies are information users, not information technology providers. Moreover, substantial aerospace companies have gone down the route of outsourcing data and information processing, and may prefer to buy archiving services. An aerospace company would still have its own archive in the OAIS sense, but that the boundary of that OIAS archive would include both the aerospace company and the archival services provider.

A software service is an ongoing provision of a capability such as storing data or migrating files to a new format. A service stack delivers a complex capability through high-level services, which build on more basic services lower down the stack. For example, to migrate a file to a new format, take it out of storage (an access service built on a storage service), do the conversion (translation service), and then store it again (submission service built on the storage service). One advantage is that basic services such as storage can be provided by a low-cost commodity service supplier, while niche services, such as CAD file translation, can be obtained through specialist suppliers when needed.

The ideal service stack is standardized and open, so that higher-level services from one supplier can be built on lower-level services from a choice of suppliers. However, as the initial development of Cloud Services showed, suppliers start by building closed, vertically integrated stacks. A lack of Cloud Service standards led to the growth of consultancies, who would select compatible services to build a complete stack. There are already signs, at least in the United Kingdom, that an archiving-as-a-service sector is emerging, focused on the lower-level services, though it is too early for interoperability standards to emerge.

This chapter uses results from Retention and Access Services in Supply Chains (RASSC), a short, UK-based project,[1] which developed an example service stack for archiving, together with cost models for information sustainment and economic and business models for the sector. This project used a scenario in which designs flowed from lower-tier suppliers up the supply chain to the prime, while feedback from the in-service phase flowed in the reverse direction.

4.2 **The Architectural Context**

The high-level concept[2] for the service architecture is summarized in Figure 4.1. The scope of the RASSC work explicitly included manufacturing and in-service data, which goes beyond LOTAR's scope.

To make the problem more concrete, Table 4.1 gives an example of some *operational performance attributes*, with different capability levels shown by example values, and with those values most applicable to the aerospace scenario shown in bold.

In this architecture, the various actors in the scenario are represented as operational nodes, which may be people or systems. Figure 4.2, the *operational node connectivity*, shows which autonomous nodes are networked to each other and send information across

FIGURE 4.1   Operational-level concept (OV-1b).

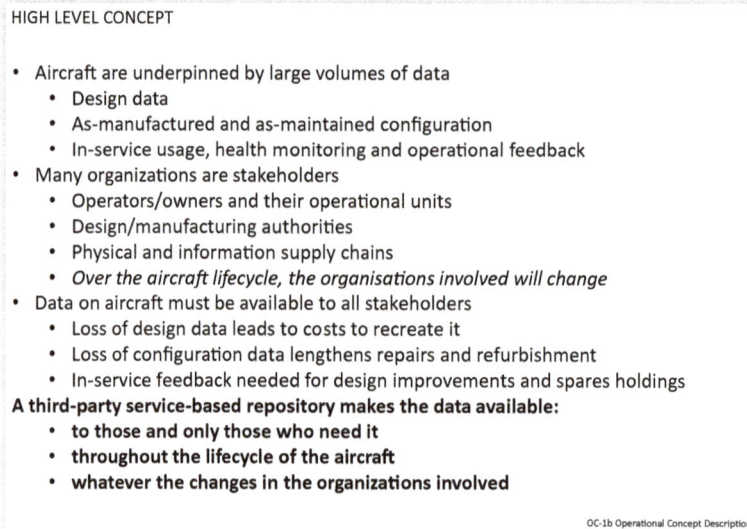

HIGH LEVEL CONCEPT

- Aircraft are underpinned by large volumes of data
  - Design data
  - As-manufactured and as-maintained configuration
  - In-service usage, health monitoring and operational feedback
- Many organizations are stakeholders
  - Operators/owners and their operational units
  - Design/manufacturing authorities
  - Physical and information supply chains
  - *Over the aircraft lifecycle, the organisations involved will change*
- Data on aircraft must be available to all stakeholders
  - Loss of design data leads to costs to recreate it
  - Loss of configuration data lengthens repairs and refurbishment
  - In-service feedback needed for design improvements and spares holdings
- **A third-party service-based repository makes the data available:**
  - **to those and only those who need it**
  - **throughout the lifecycle of the aircraft**
  - **whatever the changes in the organizations involved**

OC-1b Operational Concept Description

---

[1] Retention and Access Services in Supply Chains (RASSC), UK Technology Strategy Board project 100939, 2011 - project site now defunct, but see http://www.it-innovation.soton.ac.uk/projects/rassc, accessed March 4, 2019.

[2] Diagram numbers refer to the UK MoD Architectural Framework, a version of the US DoD's DoDAF. See [4.1].

TABLE 4.1 Operational performance attributes (OV-1c).

Attribute	Measure	Capability level		
		Low	Medium	High
Storage capacity	Petabytes	1	**1,000**	1,000,000
Storage period	Years	2–5 Short term	5–20 Medium term	**10–60 (to 100)** **Long term**
Retention regime	Information complexity level	Data (Byte stream)	**Information** **(data and semantics)**	**Knowledge** **(information + context)**
Data formats	Range of formats	Any format (at data level)	Standard formats (e.g. HTML, PDF) (at information level)	**Specialist Formats** **(e.g. CAD, F.E., etc.)** **(at information level)**
Integrity checking		Checksum	Quality checks Validation properties	**Quality, validation and** **evidential weight**
Access control		Log-on and group	**Log-on and role-based** **(local credential cache)**	**Credentials and policy** **(external credentials)**
Access times	Hours	96	1	0.01
Search capability		Fixed metadata (predefined attributes)	**Extensible metadata** **(RDF, OWL)**	**Extensible metadata** **configuration**

FIGURE 4.2 Operational node connectivity (OV-2).

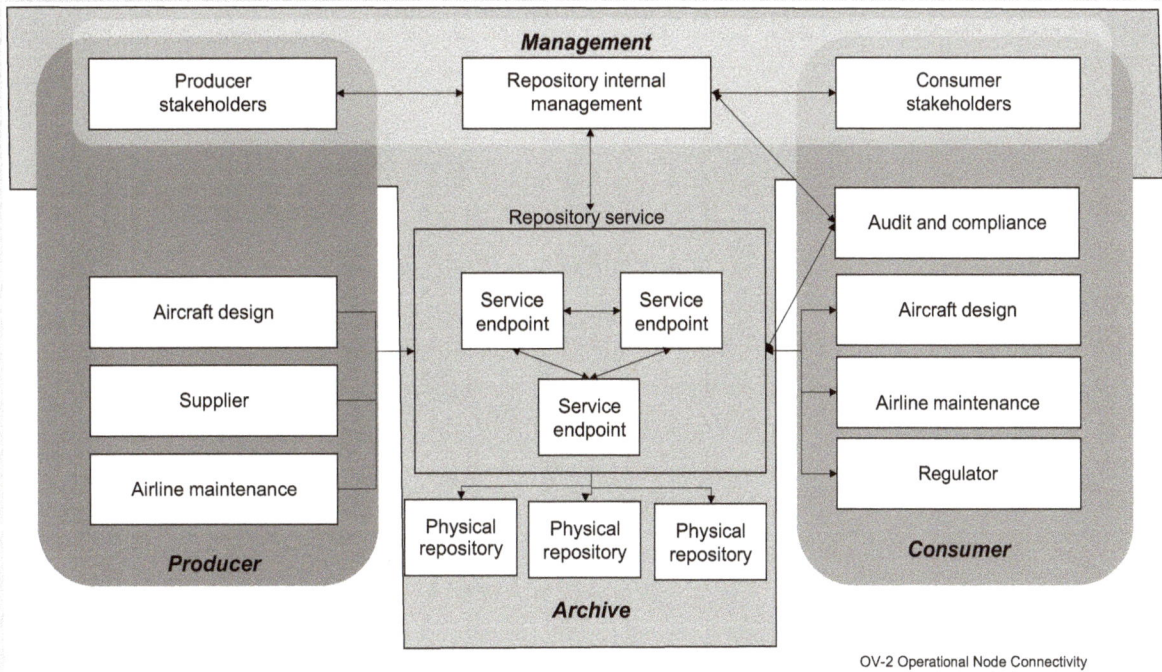

OV-2 Operational Node Connectivity

the network. It is set on a background indicating the OAIS roles of producer, consumer, archive, and archive management. This is a change of perspective from OAIS, in which each producer is implicitly independent of the others and the archive contains multiple independent collections. Here the archive has to deliver a single, integrated collection (the design of an aircraft), which is built up from the contributions of many different producers. The contrast is between filling the shelves of a library with many books and creating one very, very thick book—or at least, one per aircraft type.

However, this architecture does not require that there is a single physical archive. It assumes that major companies will want to control their own data in their own physical archive. Users access a common gateway service that provides access to multiple, federated archives, each of which may be implemented using a different service stack. For the supplier, this can mean that the information needs only to be held once in their own archive, even if they supply to multiple aircraft manufacturers. For a consumer, such as an aviation authority, a single access point provides access to all the aircraft data, irrespective of whether it is held in the prime's archive, or that of a supplier.

The *operational node connectivity* shows how the service is implemented at a particular time. This needs to be set in the broader context of how various organizations impose contractual and regulatory obligations on each other, and how those obligations are transferred over time as company structures evolve over time. The obligations are shown in Figure 4.3a. The solid lines show the contractual obligations, for example, the prime requires suppliers to make information available to the archive. In the event of industrial restructuring, these obligations are inherited by their successors—Figure 4.3b.

These relations also extended to the aircraft operators, both to supply them with the design and support information they need to operate the aircraft and also to collect operational feedback such as health monitoring data. Operational data is used by the

FIGURE 4.3 Organizational relationships—obligations (OV-4a).

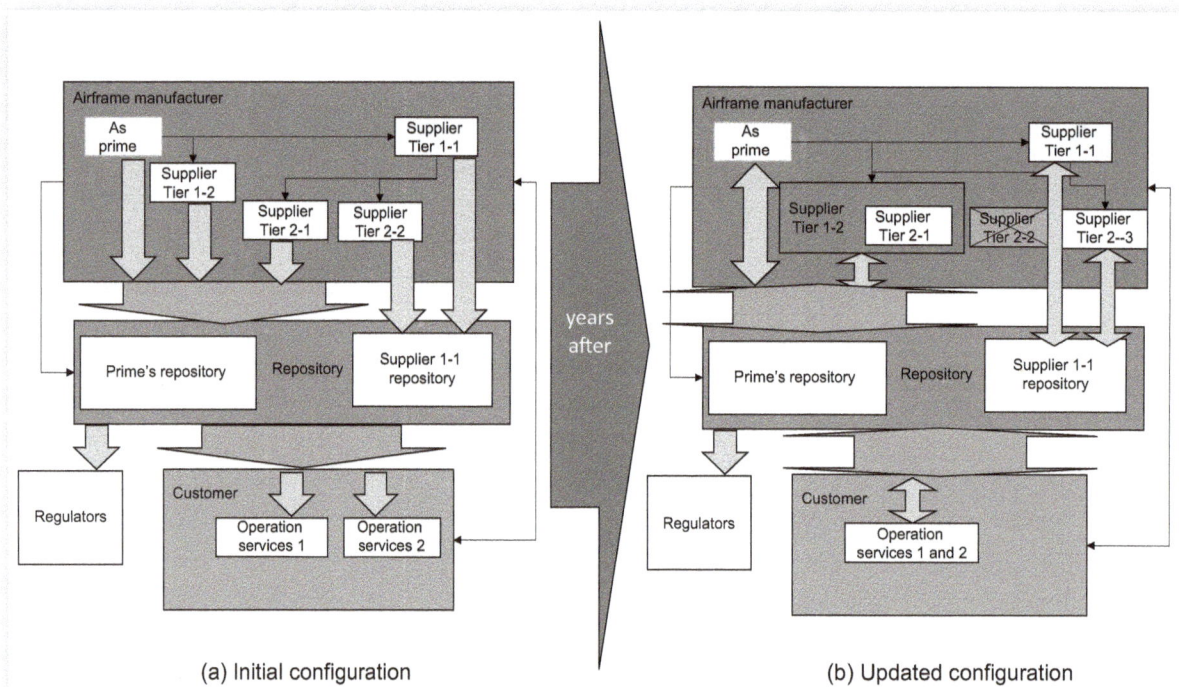

(a) Initial configuration   (b) Updated configuration

prime and the suppliers to improve both the product itself and the efficiency of the support chain. Regulators and legal organizations provide the legal and regulatory framework under which the archive operates and also access the archive. Independent auditors (not shown) certify that the archive is actually meeting the various requirements on data retention and access.

In Figure 4.3, the vertical block arrows show submission and access agreements. The primary content of these agreements covers the technical content of interactions with the archive. There will also be associated contractual agreements obliging the producers to supply information and giving the consumers the right to access information.

Should a company go out of business, there is a risk that businesses higher up the supply chain will lose access to their design information, and this could require them to rework their designs (a cost to them) or even that the aircraft loses its certificate of airworthiness. Escrow refers to information or software being held by a third party to be made available in the event that the owner can no longer fulfill the contract. Repository service providers may be contracted to hold supply chain information in escrow, although it may not be accessible to other users unless the supplier can no longer fulfill their contract.

4.3 Building Service Stacks

The architecture discussed above is predicated on there being appropriate archiving service stacks. Figure 4.4 shows a high-level view of the service stacks needed to develop and run an archive. The left-hand stack is the repository service stack, providing long-term data retention, in which producers submit, consumers access, and the archive sustains the data stored. This is discussed in more detail in the next section. The center

FIGURE 4.4 Service stack outline (SV-4 system functionality description).

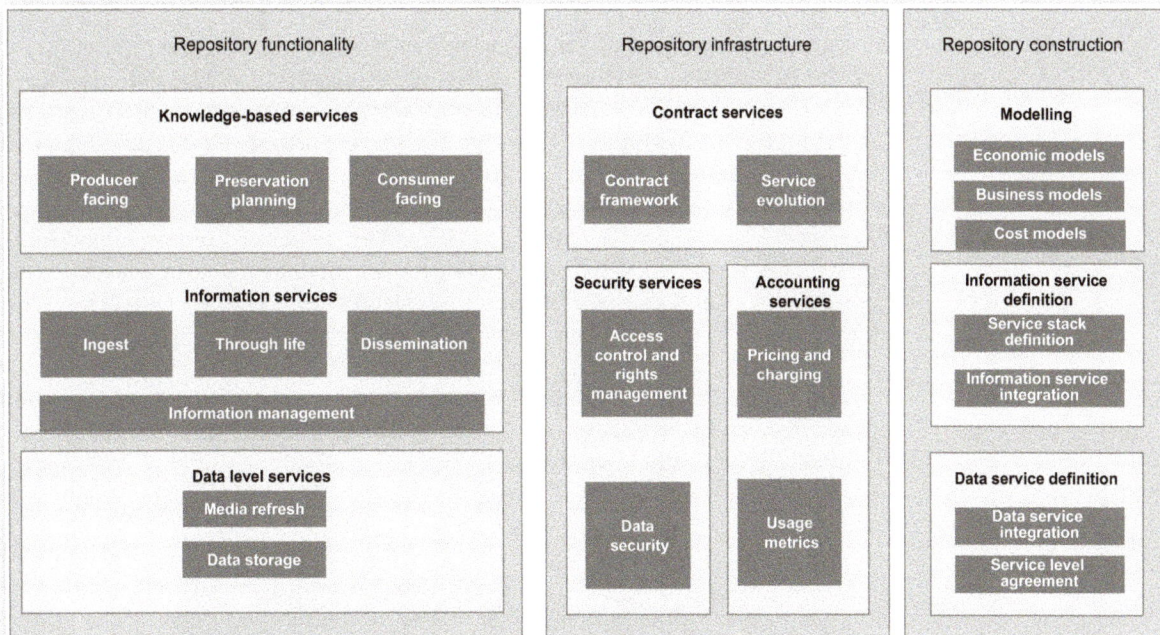

shows the infrastructure services that the archive needs to function: contract, security and accounting, and so on. The right-hand stack shows services need to construct the archive.

The archive construction services are provided by consultants and specialists, rather than being a software service. Their starting point for developing archiving-as-a-service will include cost, business, and economic models. Cost modeling provides the basis for calculating how much archiving will cost per terabyte per year, how much different kinds of sustainment service will cost, and so on—cost models are discussed further in Chapter 6.

There are various business models for a service provider, covering choices of technology, relation to the OAIS model, whether to provide just storage or to provide additional value-added services, and so on. The business models identify what market segment that the business is trying to service and sets out how a company makes money by specifying where it is positioned in the value chain [4.2]. The economic model sets the business model in a wider economic environment. For example, suppose the major aerospace companies decide to create their own vertically integrated private archives, this makes it hard for a start-up based on open-access storage services, whatever its business model. Businesses need to understand the various aspects of the economic environment, and any barriers which make particular business models hard to establish.

The infrastructure stack provides the services needed by the service provider to run the archive. For the aerospace sector, the basis for all service development is the security services, particularly access control. Access control makes access easy for those who have a valid business reason and hard for those who don't. An important use case here is where rival suppliers provide designs into the prime's shared development environment: The aircraft design team should see both designs, while the rivals are prevented from accessing each other's. It is difficult to retrofit security controls to a service stack because preexisting services may be designed with routes through to low-level services that may bypass those controls.[3]

4.4 **Archival Service Stack**

The outline repository service stack in Figure 4.4 is expanded in Figure 4.5 to show examples of individual services that the repository might provide. An important facet of this is the division of services into data-, information-, and knowledge-level services.

Data-level services require only that the service provider understands how to preserve bytes, without needing to understand the content of the data. For example, in media migration, data is copied from one medium to another, say from one type of tape drive that is becoming obsolete to a new technology. The basic function of physically copying the tape would be supplemented by verifying that the copied tape has the same sequence of bytes as the original, typically by calculating checksums. Should the drive fail to read one block of the tape, an additional service could repair the byte stream using a second copy of the data.

Basic data services could be supplemented by additional data services such as remote replication: holding the data in three or four separate locations substantially reduces the risk of data being lost to a disaster such as a fire. A question arises here: how do data replication services relate to the day-to-day operation of the business? For example, if

---

[3] For example, a Sharepoint environment had been customized to prevent the users creating subdirectories. However, the users could still create subdirectories using normal Office functionality, such as saving a PowerPoint file as a set of images.

FIGURE 4.5   Repository functionality—detail (SV-4).

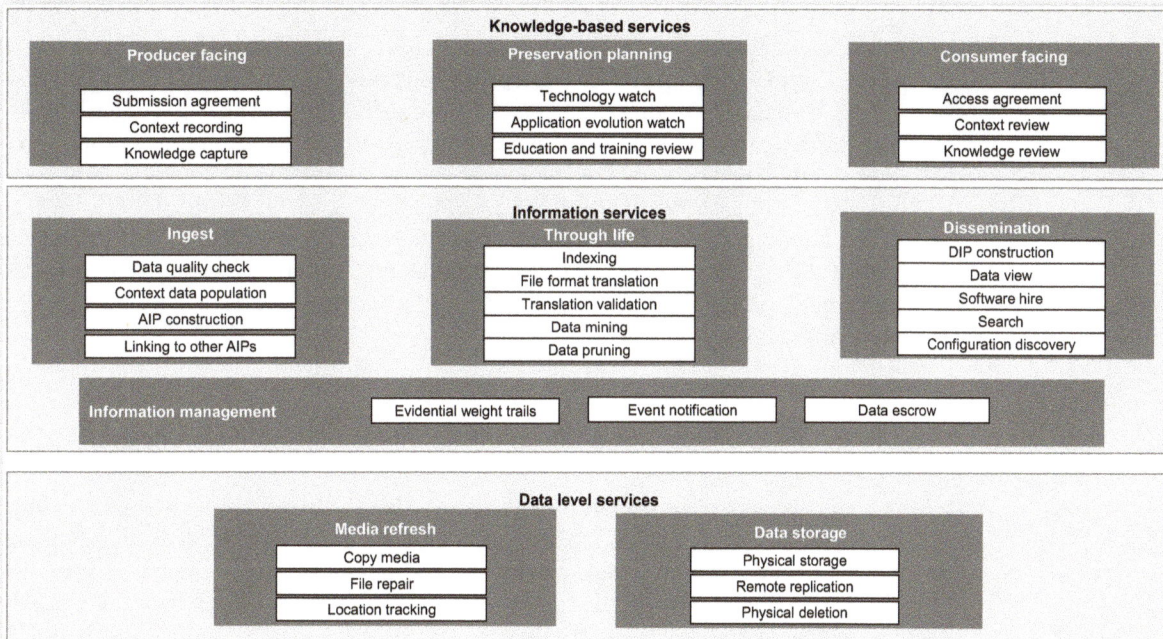

Knowledge-based services		
Producer facing	**Preservation planning**	**Consumer facing**
Submission agreement	Technology watch	Access agreement
Context recording	Application evolution watch	Context review
Knowledge capture	Education and training review	Knowledge review

Information services		
Ingest	**Through life**	**Dissemination**
Data quality check	Indexing	DIP construction
Context data population	File format translation	Data view
AIP construction	Translation validation	Software hire
Linking to other AIPs	Data mining	Search
	Data pruning	Configuration discovery

Information management | Evidential weight trails | Event notification | Data escrow

Data level services	
Media refresh	**Data storage**
Copy media	Physical storage
File repair	Remote replication
Location tracking	Physical deletion

data has a high business value, the business may require a live disaster recovery site, capable of bringing the business back online within minutes. If the repository is maintained in-house, the live recovery site may count as one of the replication sites, whereas an external archive service supplier may not have the bandwidth required were it to be used as an alternate IT site for the business. Issues such as recovery time and access bandwidth would form part of a service-level agreement (SLA), which would be cited in an access agreement.

To provide information-level services, the service supplier would need to understand the applications that interpret the data. For example, they would watch for changes to an application storage format, provide a service to translate all the old files into the new format and verify that the information content is the same. With a text document, this might involve checking the number of words was the same, checking that each word was used the same number of times, and checking that all the same fonts are available.[4] For software such as CAD, the repository may access translation software from a specialist company. It may also host the CAD software to provide remote access to consumers such as the certification authority; for example, they would have a license for the software, a service to load the CAD model into the software, and a service to share the screen with the remote user.

At the time of the original study, there was already an emerging market in data-level archive service suppliers. It was hypothesized that such archive suppliers would add additional services, moving up from the data level to the information level, and eventually to the knowledge level. It was further hypothesized that this move would first occur in

---

[4] Where a font is not available, the individual characters may be replaced by icons, characters from a different αβ, or black squares.

commodity software such as word processing or image viewing, rather than specialist areas such as CAD. At the time of writing, there is evidence of this happening [4.3].

The highest level of the service stack requires knowledge about the business and the way it uses the software. This level of service is likely to be provided by a combination of specialists in archiving and the aerospace companies who understand how their businesses use data. For example, the specialists would understand how to develop submission agreements but would need company experts to link the submission process into the aircraft design process. For commodity software, building a submission agreement may be as straightforward as selecting options in a template. But for software such as CAD, building a submission agreement needs technical expertise to understand the quirks and complexities of CAD and the quality criteria aerospace companies apply. In this context, the LOTAR standard provides relevant guidance on data formats, quality standards, and so on.

4.5 Access and Aircraft Configuration

A simplified view of an aircraft configuration is that an aircraft is built in stages—wings, body, and tail. Each stage is made up of assemblies, themselves made from subassemblies, expanding out as a "this-assembly-uses-these-parts" tree—the bill of materials (BOM)—with the leaves of the tree being individual parts (see also Chapter 10). If only aircraft configuration were that easy.

For a start, design is not a smooth progression [4.4], and as the design matures, parts and assemblies may be redesigned. In the process of manufacture, expensive parts may have minor defects and will need additional approvals if they are not to be scrapped. In operation, the aircraft may get damaged—for example an air-bridge may be docked too forcefully—and repairs will be designed and approved. Moreover, the support operation will track both parts and part locations, for example, to see if a part located in one position fails more often than those in the other locations where it is used, and this is fed back to improve the design.

So, if a designer is asked to make an update, or an investigator wants to know what has failed, there is a complex calculation to find what part is actually involved, whether the part is used in other places or has been used throughout the production run, and whether there is any history attached to the place where the part is sited. This calculation is currently done using a product data management (PDM) system. An archive's access mechanism is unlikely to reproduce the functionality of an aerospace PDM system.

This leads to three options. First, the consumer uses the aircraft manufacturer's PDM system to find the parts they need and so can explicitly identify the parts to the archive. This option is only suitable for consumers who work within the aircraft manufacturer, since only they will have access to the PDM system and have the training to use it effectively. Moreover, it is usually only applicable to the current configuration, rather than historic configurations.

In the second option, the consumer uses the archive's access service, which in turn accesses the manufacturer's PDM system. This resolves the limitation on access for consumers such as the aviation authorities and moreover, it can provide a simplified interface with a lower training requirement. Knowledge-based services for complex incremental searches have also been mooted, for example, "find all occurrences of assemblies that contain part A and also either B or C, and extract all failure reports." It still faces a possible limitation on being only applicable to current configurations, but a standardized archival interface may mitigate the knock-on effects of replacing the PDM system. Additionally, the PDM system is very unlikely to last through the whole life of the aircraft, and the manufacturer may need to migrate to a new system at some point, which may change the interface to the archive.

In the third option, the archive runs a minimal PDM viewer. This allows access to historical configurations, but will need to load the configuration before access can proceed. However, if the query is being made against the current configuration, there is a risk that the archive may not have the current working version of the data.

All of which is a long way of saying that the access services that consumers need are considerably more complicated than a simple catalog of models, but for which solutions and standards are yet to emerge.

4.6 **Sustaining Archival Services**

In the abstract, the archive sustains the information about an aircraft for the life of the type certificate; however, the physical and organizational embodiment of the archive will itself change over time. First, the infrastructure that the archive runs on—the computers, operating systems, and networking protocols—will change over time. To mitigate this, the functions of archival services can be specified from the viewpoint of the user, so such changes can, as far as practical, be hidden from the users. The greatest effort will be required if there are changes in the infrastructure that the services use to interact— for example, the network protocol—which may then entail the reimplementation of the interfaces. These must then be carefully tested to ensure that there is no change to the results of access queries: at least for legal admissibility, access queries should be demonstrably reproducible.

Second, the archive implementation may change, either by changing service stacks or by replacing the data management system. Such changes will likely entail the migration of the archived data to the new archival system. Again, access queries should be demonstrably reproducible but there would need to be a full audit of the holdings to ensure none are lost.[5] Additionally, systematic testing of a representative sample of data would be needed to ensure that the changes do not stop the data being accessed (e.g., font libraries have been carried across to the new archival system).

Third, it is likely that major companies will require control of their own data. Given the network of supply chains in aerospace, the architectural solution would be to build a federation of interoperable archival systems using a common access service. As companies restructure—merge, split, sell off business units, and so on—the individual archival systems will need to ensure that the data follows the new company structure. This may mean migrating a single physical archive to multiple distinct physical archives, or merging archives or migrating old data into the new company's physical archive. Less drastic changes will also follow, such as rewriting access permissions. Here again disruption can be minimized by federating the archival systems through a common access service. Section 6 of OAIS [3.1] on archive interoperability goes into further detail.

Fourth, contractual and legal obligations will need to be maintained. This will likely require contracts to be written in anticipation of change, including disruptive change. Such changes include the usual business evolution of mergers or business failures, but may also come from unexpected sources, such as the seizure of the data servers used because another service user is under police investigation.

---

[5] At least one aircraft company does a daily audit of its archive. Among other things, this detects computer room mistakes such as failing to mount a disk correctly, something that would not be detected by the data management system until it tries to access a file on that disk.

4.7 **Summary**

Archiving-as-a-service provides an alternative viewpoint to a monolithic architecture. It provides a way of looking at the value chain, and asking how individual services can be monetized. This leads from basic storage to added-value services such as file translation and on to knowledge-based services such as monitoring the environment, for example, to identify storage technology or file format obsolescence.

Abstracting the producer and consumer interfaces into a service definition allows implementation through multiple physical archives. By federating such archives, changes in the company structures in the aerospace supply chain could be mitigated. However, access to aerospace data is complicated by the way aerospace uses PDM to identify what parts are used in any particular aircraft.

At the time of writing, the market for service providers is immature. Although aerospace standards such as those from LOTAR provide guidance on how to sustain data, there is no common standard for archival services. It is likely that sustaining an archival service will continue to be a research and development area for some years to come.

References

4.1 "MOD Architecture Framework," https://www.gov.uk/guidance/mod-architecture-framework, accessed March 4, 2019 or Wikipedia, "MODAF," https://en.wikipedia.org/wiki/MODAF, accessed March 4, 2019.

4.2 Cave, J. (for BAE Systems), "Business and Economic Models for Supply Chain Data Repository Services," UK Technology Strategy Board Project 100939 (RASSC), 2011.

4.3 Arkivum Webinar, "Eat, Sleep, Preserve, Repeat," http://arkivum.devstars.net/webinar/webinar-recording-eat-sleep-preserve-repeat/, February 27, 2019. See also https://arkivum.com/.

4.4 Barker S, "*Aircraft as a System-of-Systems: A Business Process Approach*," Warrendale, PA: SAE International, 2018, ISBN: 978-0-7680-9402-2.

5

LOTAR: The Basics

5.1 What Does LOTAR Do?

LOTAR[1] is a long-running collaboration between many of the major US and European aerospace companies and is described on its website www.lotar-international.org. Here, we introduce the project and its aims, principally the development of the LOTAR standard (LOng-term Archiving and Retrieval of digital technical product documentation such as 3D, CAD, and PDM data).

We elaborated the project briefly in Chapter 2 when discussing requirements for archiving. Then in Chapter 3 we discussed OAIS, the general model for an archive, then Chapter 4 delved into the implementation of an archival service. The concepts from these two chapters are applicable to any types of information. With LOTAR the focus narrows to a specific aerospace concern: the archiving of design models. This chapter focuses on the general concepts of the LOTAR standard, and on the three most important topics: fundamentals and concepts, the LOTAR parts, and extensions to the OAIS processes.

In later chapters we get into the detail, using CAD as the paradigm for how to archive a design model. Chapter 7 digs into the information CAD users create so that Chapter 8 can describe the LOTAR guidance on archiving it. Then Chapter 9 runs through some of the other types of design models and signposts the work LOTAR and others are doing on archiving them. Chapter 10 covers product data management (PDM), which provides the context and metadata for design models, and LOTAR guidance on archiving PDM. Then in Chapter 11, we go beyond what LOTAR has done in PDM to look at how to scale the archiving of individual models into an archive of a complete aircraft design.

5.2 **What Information Do Designers Create?**

The starting point for a typical mechanical part—a bracket, a casing, a strut, a wing spar—is a *Work Order*, which identifies the part to be designed, the assembly where it fits, and the type of aircraft that it will be fitted to. This is recorded in the PDM system.

The shape of the part is then designed in a CAD system. The resulting 3D model is exported for analysis: for example, its response to stress may be evaluated using a finite element analysis. Once it meets all the design criteria, *product manufacturing information* (PMI), such as machining tolerances and surface finish, is added. The model is now handed over to manufacture, who may use the 3D model directly to create part programs for computer numerically controlled (CNC) manufacture: here the exact shape is important rather than the approximations used for visualization.

The part must also be integrated into its manufacturing assembly, for example, by placing the model in a virtual space to show how the whole thing fits together. A derived 3D visualization format may be used to check the process of assembly, for example, ensuring that there is an access path through the jigs used to hold the components in place. This derived format uses approximate geometry, because it requires less data and is easier to compute, making it practical to show large assemblies without the need for high-end workstations. The final design is then reviewed and approved, the *Work Order* signed off, and the PDM system updated.

LOTAR recognizes that these different models—the shape, the assembly, the part record, and the visualization—have different underlying representations. Consequently, it has developed separate guidance on the long-term sustainment of each model. Since model-based design and manufacture have been longest established in mechanical design, CAD and PDM were the initial drivers for the development of the LOTAR standard.

5.3 **LOTAR: The Project and the Standard**

In the mid-2000s, the LOTAR standard first emerged from a primeval acronym soup (see note 1 for relevant organizations and their acronyms). LOTAR the project had started earlier among a small group of European airframe and engine manufacturers and aimed to publish a standard under ASD. In parallel, from the mid-1990s the US aerospace industry was also developing a standard under the AIA, and asked PDES, Inc. to support validation and verification of the processes. The US and European teams then merged under the auspices of the IAQG in October 2002 with the first meeting in January 2003. Figure 5.1 shows the main relations.

This sort of industrial standard is created when interested organizations put together a project and share (and pay for) the work of developing the standard. Supporting organizations such as PDES, Inc. and ProSTEP iVIP have provided forums to discuss and promote the project, and recruit additional participants. The drafts are sent to the standards publishing bodies, the AIA and ASD, for quality checks before being sent out for international ballot. When approved, a LOTAR standard is then published by both ASD-STAN (as EN 9300-xxx) and the AIA (as NAS 9300-xxx—where xxx is the part number). The two versions differ only in the conventions of the publishing organizations, not in the content.

PDES and ProSTEP have been involved with the STEP standard since its inception. Since LOTAR makes extensive use of STEP, these links have been used to propose

FIGURE 5.1 LOTAR and its organizational context.

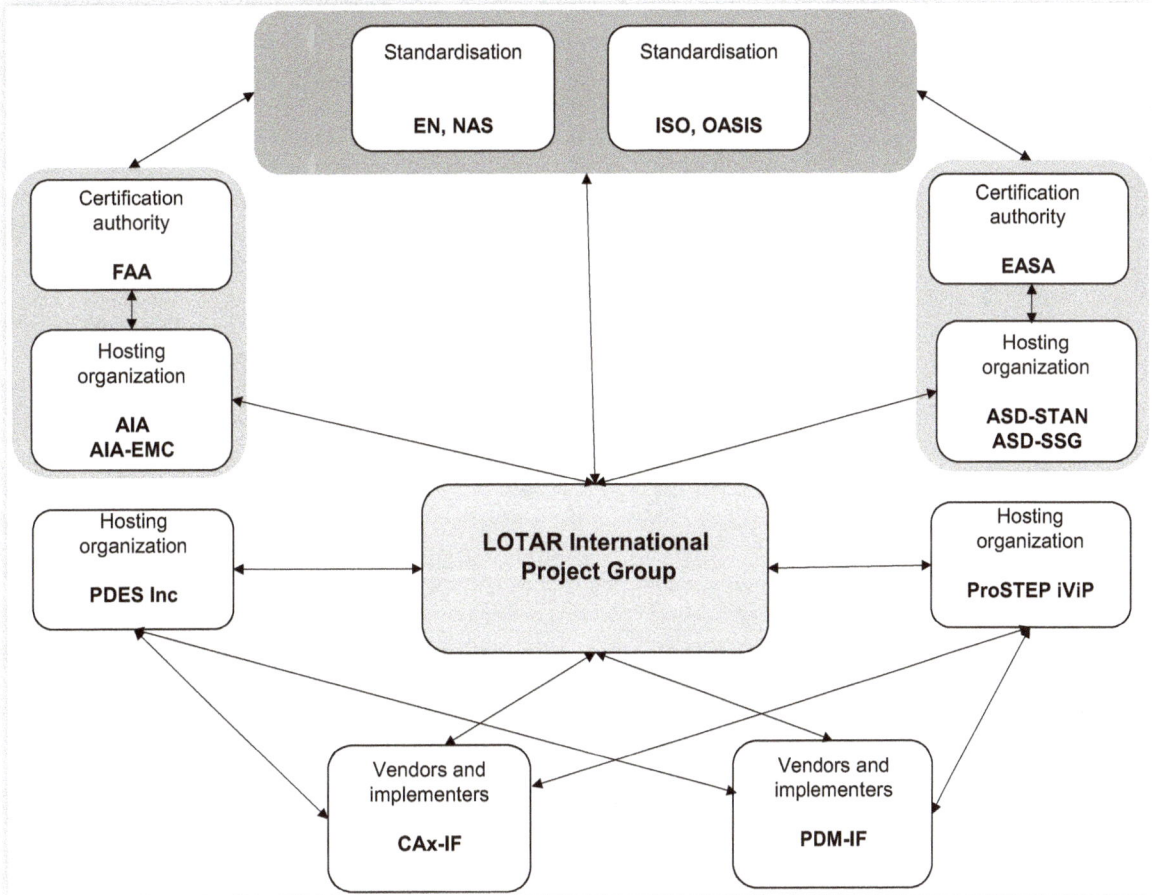

extensions to STEP and to develop a new STEP standard (AP 242—see Chapter 7). PDES and ProSTEP also host the MBx Interoperability forum, which incorporates the CAx-IF, PDM-IF, and Electrical Wiring Intercontect Systems IF. These forums test interfaces between commercial design software and also test the new features that LOTAR proposes.

Today, LOTAR boasts over a dozen members across the US and European aerospace sectors [5.2], with a further score of organizations and specialist consultancies also contributing. It has developed guidance on 3D design, covering both general mechanical CAD and more specialist applications, such as composite design and manufacture. Its scope also covers Product Structure, which is the list of parts and associated design documents that define the configuration of an aircraft and which are managed by a PDM system. The LOTAR International Project currently includes seven Technical Working Groups, which cover six domain-specific parts as well as the basic and common parts. (For LOTAR Part structure see Section 5.6.)

5.4 **What Exactly Is a Standard?**

In a technical report, options are explored and a preferred solution given, together with a rationale for the choices made. A technical standard is like a technical report but with everything except the solution taken out. A standard is normative—your solution shall do this: *shall* and *should* replace a report's *can* and *might*. Although standards may include examples, their purpose is to clarify the meaning of the normative text, not to discuss options. Where different options have been discussed, they are frequently lost to public view, continuing only in the notes of the standard-writing committee.

This narrow view of a standard could create problems for a standard on long-term archiving. In order to operate an archive, the principles behind the archive also need to be archived, and these include the LOTAR parts for the various design models. OAIS requires that this information itself would need to be independently understandable. Consequently, LOTAR also includes some parts that are primarily informative, for example, Part 2 "Requirements" or Part 3 "Fundamental Concepts."

Having a standard does not oblige anyone to follow it, nor is it the role of the standards-making bodies to enforce its provisions. However, adherence to particular standards may be required in contracts—the ISO 9000 standard for quality systems is a common requirement. They may also be imposed by regulators such as the FAA or EASA. Since aircraft operate across many jurisdictions the IAQG sponsors common standards, including LOTAR. And from the software vendor's viewpoint, standards can give a commercial advantage, because, for example, aerospace CAD users will look for a STEP interface so that they can exchange design models with their partners, customers and suppliers.

In addition to standards creators, there are often independent auditing bodies to check if an organization conforms to a process standard, or testing laboratories that check products against product standards. The CAX-IF is one model of test laboratory, where CAD and translator vendors come together to test new versions of their STEP interface.

Open standards may be contrasted with proprietary standards. Open standards are standards developed in an open process under the configuration control of independent organizations such as ISO or the AIA. A publicly available specification (PAS) uses a shorter development process, often harvesting an existing technical specification owned by a company or an industry association. For example, the 3D visualization format JT Open was originally owned and controlled by the UGS corporation; after UGS was acquired by Siemens AG, it was offered to ISO, who accept it as a PAS. At this stage, there were concerns that the format could be updated unilaterally by Siemens, but it has subsequently been republished as an open standard, ISO 14306 [5.3]. LOTAR is an open standard.

5.5 LOTAR: Fundamental Concepts

5.5.1 Core Model, Key Characteristics, and Validation Properties

An industrial standard is designed to meet a specific set of business needs. LOTAR sets out the particular needs as a series of use cases, such as "retain exact 3D geometry for reuse." A *core model* identifies the essential minimum data needed to preserve the design intent for the use case. A *key characteristic* is a feature of a material or part whose variation has a significant influence on product fit, performance, service life, or manufacturability [2.13]. The core model must therefore include the data needed to define each key characteristic. This does not imply that the data in the core model explicitly represents the information for the key characteristic. For example, if a key characteristic is the separation between two circular holes, this can be computed from the centres of the holes combined with the radius of each hole, and does not need to be recorded explicitly. This differs from the *design intent*, for example, "these holes be separated by this much to mate with part A123."

A *validation property* is a property derived from the model, which is used to check that it has been preserved correctly. Ideally, the set of validation properties for a model would ensure any change in any key characteristic that would be identified. In practice, the focus has been on general validation properties, such as the volume and the centroid of a 3D model, or on additional validation of important properties, such as the position of points along the edges of a fan blade. Passing such a validation test does not prove that the part is unchanged; however, properties are chosen to minimize the probability that an altered part could pass such a test. Validation properties, such as volume, became established as checks on exchanges between organizations, although in practice it has been found that many failures can be related to poor-quality models, rather than the exchange process itself.

A validation property is computed both in the originating software and in the software that is used to read a part taken from the archive. This implies that both software systems should be able to compute the property and in the same way; for example, a volume calculation should account for any voids internal to the part in the same way. Use of validation properties ideally requires that the validation property is archived with the part. In practice, in STEP the standard has been extended to include validation properties, so these can be saved in the STEP file.

5.5.2 Model Representation

A model is a combination of data that defines the static form of the model and the software that interprets the data, and which may create a working form for use in computation. One analogy has the data as the ingredients, and the software as the recipe—both are needed to create a cake (a particular CAD model). The data is said to be a representation of the model, and the particular representation that the software usually stores is called the native representation. An alternative to a native representation is a neutral representation such as STEP. Choosing the right representation is a key decision in archiving.

Software changes over its life—it goes though versions and generations. Versions are the regular updates, mostly providing bug fixes and incremental improvements. A new generation is a substantial change, and often the native form is altered to match the new software. Often, this requires the user to actively migrate from one generation to the next. The various generations of Microsoft Word illustrate the changes in file formats, with older versions of Word documents being transformed when read into later generations of Word—which is why PDF is often cited as a preferred archive format for documents. For LOTAR, long-term means longer than a software generation, while short term covers several software versions within the same generation.

Although the native format may change between generations, the neutral form usually will not. Moreover, since the neutral form is defined in an open standard, the documentation is available to reconstruct the software needed to read it, whereas some software vendors restrict the availability of documentation on the native form.

A neutral form maybe also be classified as a derived form, that is, it is derived from the native format. The distinction is important because going from one to the other requires a transformation of the data, and that requires validation to show the two forms represent the same model. LOTAR requires both that such a validation is performed and that the results added to the audit trail [L11]. Going from the native form to the neutral form may also result in data loss if the native form includes proprietary constructs, which are out of the scope of the standard. Some companies [5.4] require that their designers only use constructs that can be exported to STEP, which means that the model can be archived in STEP form without loss of any feature.

Besides classifying the stored form as native or neutral, for CAD the stored form may be detailed or approximate. In CAD, a detailed form provides the exact geometry, whereas an approximate one simplifies the geometry, for example by replacing a smooth surface with one using polygonal facets. An approximate form is computationally less expensive to use, but the approximations may reduce the quality of some operations, such as machining a smooth surface.

By contrast to a representation, a presentation is a derived visualization, such as a screenshot. Presentations do not preserve the model. Although it is easy to print out a presentation, looking on edge at the paper confirms that the three dimensions of the original are lost. Note that 3D visualization formats generally use a simplified representation, because operations such as 3D rotation cannot be performed on presentations such as 2D images.

5.5.3 Quality, Validation, and Verification

LOTAR identifies three sorts of checks for archived models:

- A quality check is applied to the original model to ensure it is fit to archive
- A verification test ensures that the model is correctly represented in the data
- A validation test ensures that the model is unchanged when read into new software

Quality failures range from basic errors, such as not actually saving the 3D model, to more complex ones, such as creating the same hole in the same place twice. In the latter case, the model appears visually correct, but the hole has two distinct entities as its inner surface, which risks the model being rejected in the retrieving software. A verification is a type of quality test that ensures the model has the correct structure; for example, there are no gaps between different surfaces of a solid.

A validation test ensures that key characteristics are unaltered; for example, a point on a given surface is still on the surface when read in by different software. A validation criterion works at the level of information—how the data is interpreted by software— rather than at data level (e.g., the bits are unchanged).

Many of these tests can be implemented through commercial model quality checkers. Such software may also implement ways of repairing invalid models, for example, surface healing modifies models to remove gaps between faces. LOTAR mandates quality checks, validation, and verification, but model repair after reading from an archive may invalidate some purposes such as legal admissibility.

5.5.4 Digital Signatures

A digital signature is used to ensure a model is not changed while it is stored, or while it transferred between the producer and the archive or the archive and the consumer. A digital time signature is used within the archive to ensure that the stored data has not been altered. LOTAR also notes the engineering signature as a subtype of digital signature; this is attached to a model to say that it has been approved for use in an aircraft. Because of concerns about the need to renew digital signatures, LOTAR only uses digital signatures to confirm the transfer between the producer and the archive.

5.6 LOTAR: Parts of the Standard

With model-based engineering continuing to be taken up by new disciplines, it was always anticipated that creating a comprehensive standard for aerospace would be a long job. Consequently, LOTAR is published as a series of parts. To navigate the standard, they have been divided into groups (Figure 5.2)[2]

FIGURE 5.2 Structure of LOTAR parts.

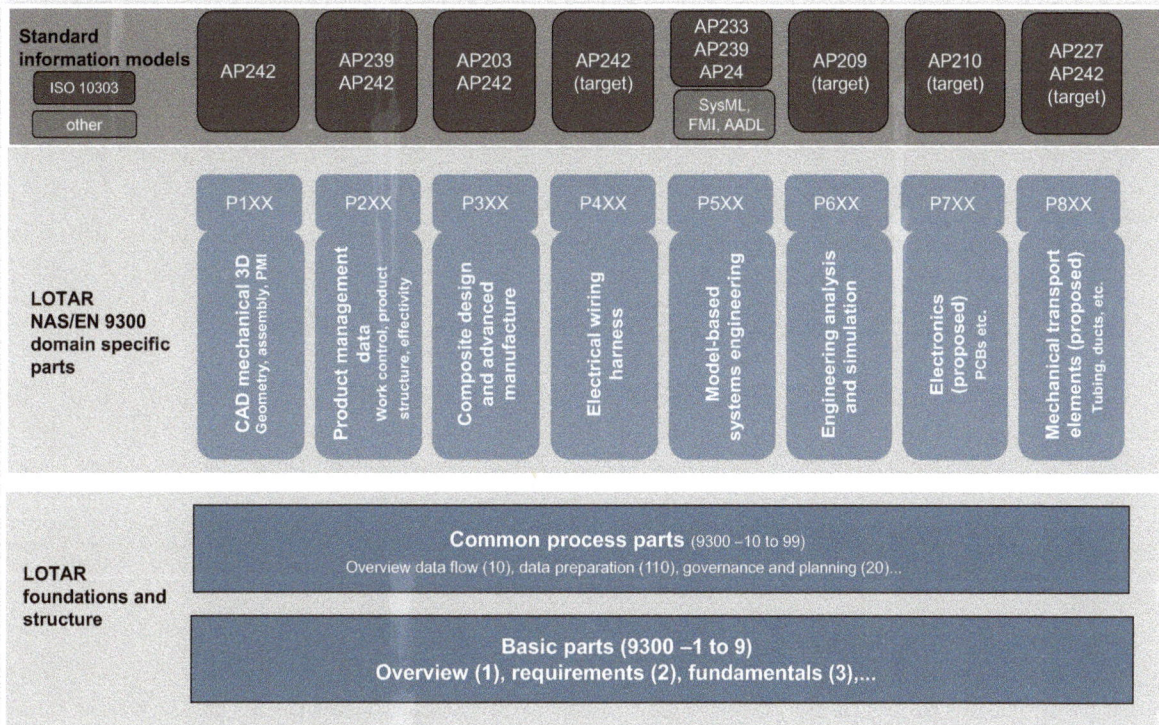

Standard information models ISO 10303 / other	AP242	AP239 AP242	AP203 AP242	AP242 (target)	AP233 AP239 AP24 SysML, FMI, AADL	AP209 (target)	AP210 (target)	AP227 AP242 (target)
	P1XX	P2XX	P3XX	P4XX	P5XX	P6XX	P7XX	P8XX
LOTAR NAS/EN 9300 domain specific parts	CAD mechanical 3D Geometry, assembly, PMI	Product management data Work control, product structure, effectivity	Composite design and advanced manufacture	Electrical wiring harness	Model-based systems engineering	Engineering analysis and simulation	Electronics (proposed) PCBs etc.	Mechanical transport elements (proposed) Tubing, ducts, etc.

LOTAR foundations and structure

Common process parts (9300 –10 to 99)
Overview data flow (10), data preparation (110), governance and planning (20)...

Basic parts (9300 –1 to 9)
Overview (1), requirements (2), fundamentals (3),...

The main content of LOTAR is the detail of archiving design models, which is found in the data domain-specific parts. These provide criteria, which minimize the risk that a model will not be recoverable in the future and ensure that if the models are read incorrectly, then those errors will be detected. Each domain is allocated a number range, for example, the 100s for 3D CAD and the 200s for PDM.

Within each domain, a number of use cases are identified, each describing what model access is needed in a particular circumstance. To meet each use case, a core model is delimited, identifying the essential minimum data needed to preserve the design intent for the use case. Typically, the core models for each use case overlap, so that the first part in any domain covers the smallest core model, and so the next part covers a larger scope, and so on.

The standard tries to avoid duplication between core models, so some use cases map to multiple core models (Figure 5.3). For example, to capture a 3D assembly, one needs a 3D geometrical model of part, a product structure for the assembly (the configuration of parts it contains), and a 3D assembly model in which the 3D parts are aligned to each other.

Each domain part therefore specifies the information in its core model, together with data quality checks needed when the data is prepared, and also validation and verification criteria that apply. Chapters 7 though to 11 get into more detail on domain parts.

The data domain-specific parts are supported by the basic parts (numbered between 1 and 9), which record the basic principles behind the standard, and the common parts (numbered between 10 to 99), which detail LOTAR guidance on an implementation of an archive.

FIGURE 5.3   Use cases, core models, and parts.

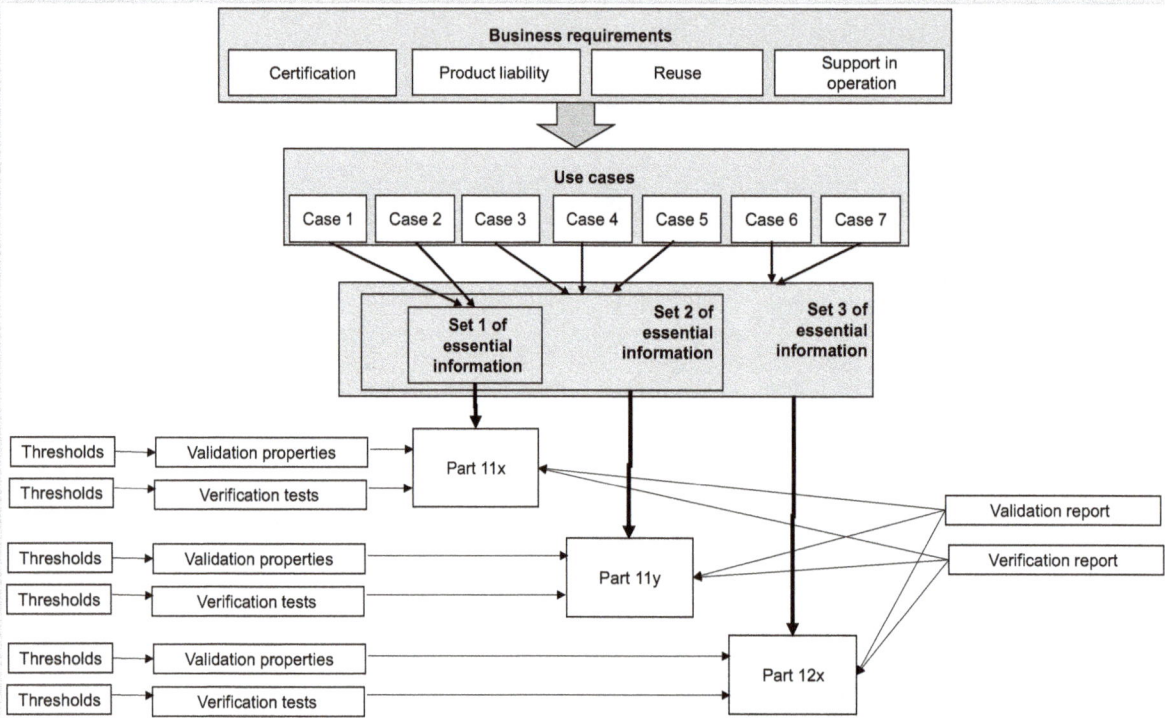

5.7 LOTAR: Processes and Extending OAIS

LOTAR parts 10 to 15 detail the archival processes, extending the OAIS processes of Ingest (part 12), Archival Storage (part 13), and Retrieval (part 14), but also adding processes for Data Creation (part 11) and Removal (part 15).

Much of the detail within the LOTAR processes is concerned with audit reports for particular steps in the process and with error handling. Data Creation—which creates the Submission Information Package (SIP)—and Ingest—which creates the Archival Information Package (AIP) from the SIP—have quality checks added, and the audit reports from these checks are added to the SIP or the AIP. In the process of Archival Storage, a digital time stamp is added to the AIP to demonstrate that it remains unaltered while it is in storage.

A key step in the design process is approval of the design model as being fit-for-purpose, and this is also the point when the aircraft configuration can be updated to include the new design. Approval is a relatively expensive process, since it involves experts from multiple disciplines checking the model against the design rules applicable to their discipline. For example, among the checks on simple metal parts, there will be a check that the electrical paths it creates will not compromise the aircraft if it is struck by lightning. Consequently, when integrating the LOTAR Data Creation process into the design process, it is advisable to go beyond the LOTAR process and put the archiving quality checks before approval (Figure 5.4b). Quality checks could be done after the approval (Figure 5.4a), but if the checks failed then the model will need updating and the approval repeated. For CAD parts, many of the archival

FIGURE 5.4 Approval and archiving.

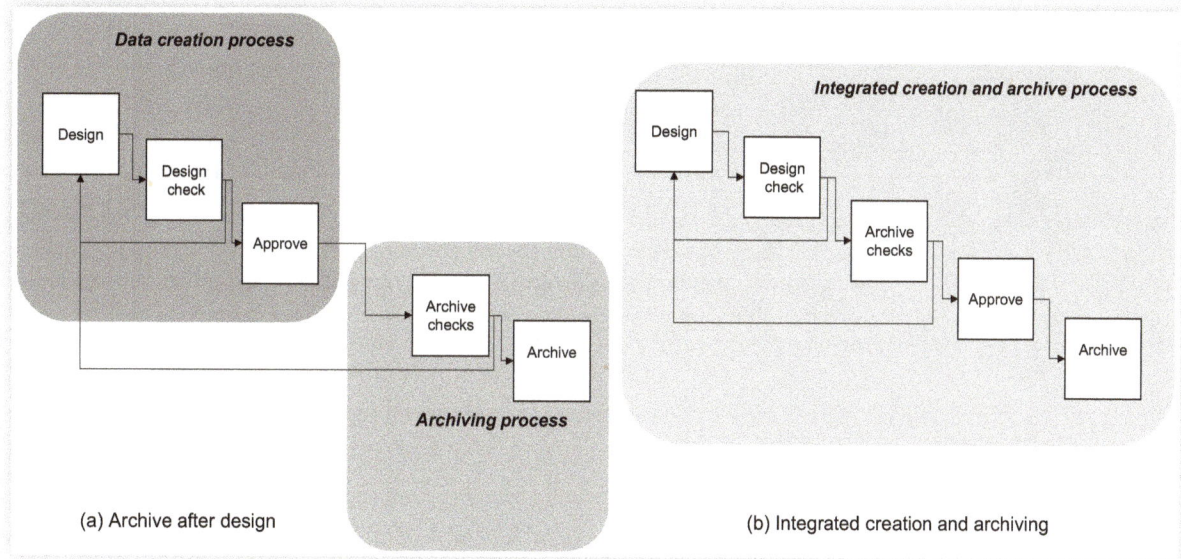

(a) Archive after design

(b) Integrated creation and archiving

quality checks are best done automatically, since modelling errors may not be visible on visual inspection. For example, a shape may be accidentally duplicated over the top of the original.

An implication of *check-fit-for-archive* before *approval* is that the archive should be in place at latest by the start of the final design phase. There is a temptation to leave the cost of archiving until later in the project, because the design software should remain usable for several years. However, if that is done, then the project will run the risk that models may fail quality checks and have to be updated to archive them. Users—or user management—have a tendency to say "archive the existing data anyway, and we will behave better in future," which is rather like telling someone to file away the contents of the rubbish bin. If poor quality data is archived, it may prove unusable to later generations of software. However, the design of the aircraft prototypes may not need archiving, as prototypes generally have a much shorter life cycle.

5.8 **Summary**

The LOTAR standard is divided into parts. The initial series of parts (1 to 99) covers general archiving principles, but the main content is in the domain-specific parts (100 onwards). Each domain-specific part starts from a use case in a particular design discipline. It defines the key characteristics that must be preserved, the core model that can represent those characteristics, and validation checks to show that the characteristics are preserved unchanged.

LOTAR also extends the OAIS processes to include quality checks and to add the audit trail from those checks into the archive.

The LOTAR consortium sits in a broader context of organizations that publish and promote standards, and aerospace-specific organizations that mandate standards for aerospace.

References

5.1   PDES, Inc., "Strategy for Long Term Product Data Retention Using STEP," Jan 1996.

5.2   LOTAR—Member Organizations, http://www.lotar-international.org/lotar-organization/member-companies.html, accessed February 19, 2019.

5.3   Wikipedia, "JT (Visualization Format)," https://en.wikipedia.org/wiki/JT_(visualization_format), accessed February 27, 2019.

5.4   "How Gulfstream Ensures Data Integrity for MBD," https://www.engineersrule.com/gulfstream-ensures-data-integrity-mbd, accessed January 14, 2019.

---

[1] There are many organizations involved in the development and publication of standards, and LOTAR works with just a few:

- Standards setting bodies

  - AIA—Aerospace Industries Association—US-based publishing the National Aerospace Standards (NAS prefix)

  - ASD—AeroSpace and Defence Industries Association of Europe—European-based publishing standards with the EN prefix

  - ISO—International Organization for Standardization (ISO from the Greek for the same, because, for example, in French an acronym would come out as OIN)

    - TC184/SC4—Technical Committee 184/Subcommittee 4—Industrial Data Standards, the ISO subcommittee responsible for the STEP standard

      - STEP—STandard for the Exchange of Product model data: ISO 10303

    - SAE International—standards for aerospace with the AS prefix

- Supporting Organizations

  - PDES, Inc.—US industry-based organization supporting model-based development and manufacture with strong links to the aerospace sector, and with a strong input to STEP development

  - ProSTEP iVIP—European industry-based organization for digital transformation in design and manufacture, with strong links to the automotive and aerospace sectors, also with a track record in STEP development

  - ProSTEP AG—a consultancy company independent from but linked to ProSTEP iVIP

- Other organizations

  - IAQG—International Aerospace Quality Group, coordinating quality standards development and deployment across aerospace

  - MBx-IF https://www.cax-if.org/index.php (accessed November 25, 2019), which incorporates:

    - CAX-IF—the CAD software developers Implementers Forum, which does compatibility testing between STEP interfaces and tests out new STEP features

    - PDM-IF—testing interoperability between PDM vendors

    - EWIS-IF—Electrical Wiring Interconnect System

[2] 5.6] LOTAR Parts list from http://www.lotar-international.org, "The LOTAR Standard" and various progress reports. Accessed February 20, 2019.

1	Structure
2	Requirements
3	Fundamentals and Concepts
4	Description Methods
5	Authentication and Verification
7	Terms and References
10	Overview Data Flow
11	Data Preparation
12	Ingest
13	Archival Storage
14	Retrieval
15	Removal
*20	Governance and Preservation Planning
100	Common concepts for long-term archiving and retrieval of CAD 3D mechanical information
110	CAD mechanical 3D Explicit geometry information
115	Explicit Assembly Structure
*120	CAD 3D Explicit Geometry with Presentation of Product and Manufacturing Information
*121	Semantic representation of CAD 3D Explicit Geometry with Product and Manufacturing Information
*125	Explicit CAD assembly structure with graphic presentation of Product and Manufacturing Information
200	Fundamentals and Concepts for long-term Archiving and Retrieval of Products Structure Information

* Proposed for publication

CHAPTER 5

Governance, Planning, and Preservation Planning

6.1 Governance: Getting the Archive You Want

Governance translates the aims of the business into goals for the archive. For example, the business aims to deliver the next aircraft type for ten years and then support for forty years: the archiving goal is to sustain the design over the next sixty years—add a period at the beginning when the design is being created, and another at the end for product liability to expire. Planning the archive takes the goals and works out how to deliver them: what needs to be done, how long it will take, what it will cost, and what are the residual risks. Preservation planning monitors those residual risks in order to reduce them. Those risks include technical (e.g., storage technology is becoming obsolete), commercial (e.g., is our software supplier going bust?), and the users—is the designated community losing the knowledge needed to understand archived data?

The model of governance proposed here uses a cascade of four levels of competence, from the executive level, which defines the business strategy, through engineers and technical specialists, down to the IT department who must enact the archiving plan. These competences are based on the knowledge each level maintains: the executives understand the airline customers and their needs, while the IT department knows how to replace a disk drive. A major driver for ongoing planning is the feedback loop between technical specialists and engineers, which identifies risks such as formats becoming obsolete or the company issuing software licenses going out of business. This model is based on early drafts of the LOTAR standard on Governance and Preservation Planning [L20].

To give further depth to governance, three topics are discussed in more detail. An archiving cost model is needed to set the cost of the archive against the gains from sustaining the design. Provenance—who created and who owns a model—is important both for understand where the costs lie and for demonstrating the legal admissibility. Audit is required, both of the archive to show it is working and of the archive service supplier to show they can deliver a long-term service.

The Governance and Planning model overlaps with the OAIS archival management function, but, being specific to an industry, provides more detail in some areas, particularly the different layers of governance.

6.2 **Layers of Governance**

The model of governance (Figure 6.1) separates four levels of knowledge and authority layered on top of each other:

- Business level—senior executives define retention policies based on plans for the business and the levels of business risk that can be accepted.

- Engineering level—to meet the policies from the business level, senior engineers develop use cases that identify the access needs of the designated communities.

- Technical level—specialists in topics straddling the engineering/computing boundary, such as CAD data formats, identify preferred archiving methods for the engineering use cases.

- IT level—provides the services to store design data according to the methods mandated by the technical level and maintains them, updating computing technologies when needed.

These levels of knowledge must be mapped on to the physical organization. For example, where IT has been outsourced, the IT level will be provided by the

FIGURE 6.1   Governance model.

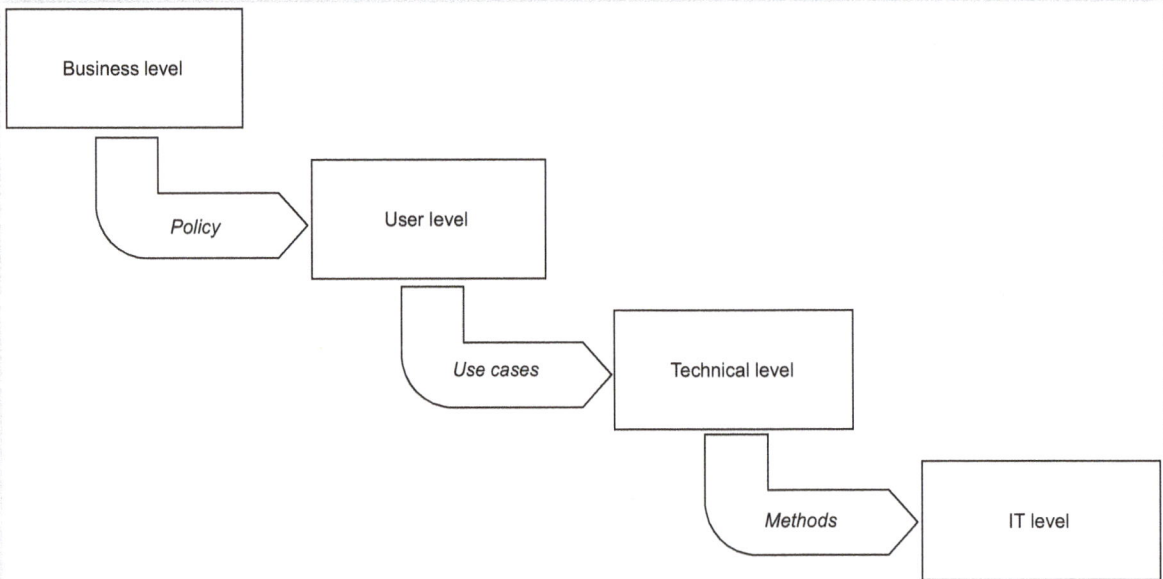

outsourcing contractor. In terms of the service model described in Chapter 4, the outsourcer will provide the data-level services. Some outsourcing contracts may extend to the technical governance level and may provide information and even some knowledge services—although in many major aircraft manufacturers, understanding of CAD formats is retained as a core skill. In small supplier organizations, much of the work below the business level may devolve to a single person, and their main strategy may be to follow the requirements and guidance of larger companies higher up the supply chain.

6.2.1 Business Level

The main drivers at the business level will be:

- Complying with regulations on information retention.
- Risks should information be lost.
- The strategy for future business, particularly planned revenue streams such as sales, spares, and after-sales service.

The output will be policies that control those risks, for example, that engineering should (a) plan to give regulators access for a minimum of 30 years and (b) advise on the costs of access for a further 30 years.

By way of counterexample, if the executives assumed that the company could close in 5 years, then they would ignore long-term retention, store design data in a bucket of sequentially numbered PDF files, and take bonuses for the reduced running costs. Beyond 5 years, the executives would have moved on but the customers and investors would lose out when the aircraft loses its type certificate. Such short-termism might be avoided were it is mandatory to publicly report the company's data retention strategy; however, as the Enron scandal suggests, regulation tends to follow after bad practice [6.1].

The other major inputs to policy development are the costs of running an archive and audit reports showing that the policies have been implemented—or showing where the implementations are falling short.

6.2.2 Engineering Level

Engineers will know what information they create, which of that information they will need in the future, and how they will want to access it. These are formally described in use cases, to be passed down to the technical level. Use cases situate the information in the wider context in which it is understood—here that context is the designated communities for the archive. However, these use cases also require a leap of imagination to understand what an engineer 20 or 50 years in the future will need. That leap of imagination needs to consider three basics: the user environment, the user community, and quality criteria for the information.

The user environment describes how the user will access the information and is both the most predictable and least certain. It is most predictable in that accessing the data will need software and hardware, and that although technologies will change, the seeds of future technologies take time to germinate, often taking 20 years or more from initial proposal to a commercial system—we are still waiting for the gesture-controlled, holographic CAD screen. Where it is least certain is that IT suppliers come and go, often quite suddenly, which makes the use of proprietary software and data formats particularly risky. Consequently, the engineering level needs to monitor the commercial environment as well as the technical environment and ensure changes they note are passed down to the technical level.

The user community brings with it a particular body of knowledge, and that knowledge will be implicit in a use case. Preservation planning requires that the body of knowledge is explicitly described (see, for example, the Systems Engineering Body of Knowledge [6.2]). Although there are occasionally radical shifts in the way things are done—for example, the move from analogue to digital electronics—mostly there is a slow, continuous change. Eventually this will leave new entrants in the field without the knowledge of how things used to be done or the skills to do them. For example, a key design office skill used to be interpreting a three-view drawing as a 3D shape; however, with 3D visualization software becoming the norm, there is a risk this skill will fall into disuse, and as a result, old drawings may become practically unusable. One way of retaining such knowledge is to maintain a library of old textbooks, so that a younger engineer can work back to the sources. One way of losing such knowledge is "storage rationalization," with the furniture police[1] forcing a clear out of documents that are not immediately useful.

The quality criteria are set in terms of "key characteristics"—the details which any archive must sustain, as opposed to the accidentals such as paper color, where changes do not lose the intent of the designers. These will depend on the engineering domain. It is also tricky to express the difference between the key characteristics that identify engineering intent and similar looking details, which can be derived from the design. For example, a slot may be designed to be a fixed distance from the left edge of the part, but the distance to the right edge will follow automatically from the size of the part and may not itself have any particular design significance. Formulating such key criteria equates to formalizing engineering knowledge.

6.2.3 Technical Level

The technology of information representation is a knowledge base distinct from that needed by users. The key questions are whether the software represents the user information adequately and accurately, and whether any change in the format retains or distorts that information. An example of this comes from the 1990s when design processes switched from being drawing-based to model-based (see Chapter 10). Sharing configuration management required a switch from the AECMA message to STEP format,[2] although, in this case, management focus was more on the £13,000,000 bill for replacing IT systems.

The technical expert role is not that of a software developer—although in practice most have written software. For preservation planning, the technical expert must be closely connected to the engineering processes and understand how change impacts the way data is sustained. Since business change may also involve migration to new versions of software, or even to completely different software, the role must also be able to interpret information structures and be able to identify the significance of any changes. In particular, the technical level should identify formats and standards that preserve the required key characteristics and which preferably include validation properties to check the key characteristics have been correctly retrieved.

---

[1] For "furniture police," see [6.3].
[2] Reported by the author at a PDES Inc. meeting c. 2002.

6.3 **How Much Will It Cost?**

A cost model is a model used to predict the cost of a particular product, process, or service. For archiving, there are broadly three types of cost model [6.4]:

- Empirical, where the preservation lifecycle is broken down into functional units small enough to make a cost estimate for the unit.

- Historic, where costs are extrapolated from experience.

- Simulation of the repository, based on the services, processes, and resources consumed.

However, once one gets into more detail, cost models for archiving diverge, both in their structure and the results they give. These models have been studied in the libraries and academic sectors, notably through the 4C project and the follow-up Curation Costs Exchange.[3] These focus more on the retention of documentation,[4] so in some aspects— particularly the need for validation properties—their applicability to sustaining design models is limited.

To give some flesh to cost modeling, an empirical model is outlined below, simplified here from [6.4]. This breaks the costs into Submission, Storage, and Dissemination costs. If T is the timescale over which the costs are calculated, then

a. Submission costs are made up from:

Submission workflow cost $=$ [Fixed cost of creating each workflow]
$+$ sum over each workflow of [cost per submission] $*$ [rate of submission] $*$ T

Producer support costs $=$ [Fixed cost of support]
$+$ [cost per user] $*$ [number of users]

b. Dissemination costs are similarly made up from workflow and consumer-support costs.

c. Costs of storage and preservation are made up from four elements:

Content Type Costs $=$ [Fixed cost of support]
$+$ [cost of a new content type] $*$ [number of types]

Storage Costs $=$ [Fixed cost]
$+$ sum over every submission workflow of [unit cost of storage]
$*$ [mean storage volume per submission] $*$ [rate of submission] $*$ T

Storage Ingest Costs and Storage Access Costs
$-$ similar formulae to that of storage costs.

---

[3] "Collaboration to Clarify the Costs of Curation," http://www.4cproject.eu/. The original project ran to 2015, but has been followed up by the Curation Costs Exchange, http://www.curationexchange.org/, all accessed May 13, 2019.

[4] *External reviewer comment* They look at digital assets quite generally, but principles are most clearly worked out for repositories of research data (encompassing not only textual documents and spreadsheets/tables but also audiovisual material, executable workflows, more complex formats from statistical and qualitative analysis packages, databases, stereolithography, etc.). The problem is not so much with the type of asset under consideration, but that these projects are catering for a kind of lowest common denominator across the academic sector, rather than the specific needs of a given community or the difficulties peculiar to a family of digital asset types.

This cost model is more applicable in short-term data center costs for simpler formats such as documents or aircraft health data, but gives limited insight into the costs for, say, CAD models. For CAD, first, the submission and dissemination costs need to be modeled in more detail to include the costs of quality checks, creating validation properties, and the cost of maintaining audit trails. Second, the need to move files to a new format is unpredictable, so the cost of transforming the files is not included in the storage and preservation costs. Third, costs for preservation planning are not included, particularly the costs of monitoring the environment for business and technology change, or for changes in the designated community. Moreover, the costs of file format transformation and preservation planning are likely to see step increases over time, as gap between the original technology and the current environment increases.

A further complication is that different preservation approaches result in different risks, and planning the sustainment strategy involves trading the costs incurred if the worst happens against the costs of mitigating that risk. For example, consider the risks that the repository is destroyed by fire, flood, or earthquake. These events are individually unlikely, but the cost of losing all the product documentation is enormous, not only in terms of regenerating the data, but in terms of loss of reputation. This risk can be mitigated by creating a second repository in a different location. This will double the cost of the storage element of long-term archiving, but will not affect the costs of creating submission agreements.

There is, of course, the chance that both repositories will be lost at the same time (don't site both repositories on the San Andreas fault). Here the risk can be further reduced by having a third repository, or even a fourth, preferably widely separated with a least one on a different continent. At some point, the decision must be made that the reduced risk from additional repositories is not worth the extra cost.

Each governance level makes its own contribution to cost drivers, with the business level defining the acceptable level of risk, and the IT level providing historic costs of storage, processing, and media refresh.

6.4 Provenance: Trusting the Data

An implicit requirement for an archive is that the data remain trustworthy. Provenance tracks who created or modified data, and when and in what context. Provenance also identifies who owns what intellectual property—important during company restructuring or transfer of design authority.

In aerospace there are two main strands to provenance: information recorded in Product Data Management systems and data arising through the archiving processes. In the latter strand, an audit trail links the submission information package (SIP) through its various transformations into AIPs and DIPs, and provides evidential weight that links the DIP to the original submission, showing that the data retrieved is the data archived.

In aerospace, Product Data Management (PDM)—the other strand of provenance—has the particular significance of providing the process audit trail running from design through to certification. The PDM system will record provenance data such as who created the model, through what process, and in particular, how the model was checked and approved, who were the checkers and approvers, as well as the aircraft configuration it is approved for.

Preservation planning has the role of identifying the provenance requirements for the various designated communities, and ensuring the processes deliver that information. In particular, at the technical and IT levels, it will need to consider how PDM information is linked to the models archived. This is discussed further in Chapter 11.

6.5 **Audit: Checking the Archive Works**

There are two senses for audit relevant here: showing the current repository is doing what it is supposed to and showing that the archival system continues to be capable of long-term information sustainment.

The first type of audit lists the archive contents to show that the archive contains all the information that it is supposed to. This is a test that the IT department has not accidentally lost a disk or renamed it so it is no longer accessible. Regular checks avoid the risk that the people who made the mistake have moved on, since they are most likely to know where the missing data is and can restore it. Additionally, there is likely to be a specific repository for approved data, where files that have not been authorized must be found and removed, and the process failures which let them in must be corrected.

This type of audit also involves accessing sample models to demonstrate that they can actually be retrieved. The samples will cover each use case, and tests should use the access methods available to each designated community. For example, if the aviation authority uses remote access to view a CAD file, the remote access must be tested in a way that mimics their computer environment, including the software license keys they have and the security controls they need to unlock.

The second type of audit—that the archival system is itself sustainable—is described in ISO 16363 the Trusted Digital Repository (TDR) Checklist [6.5], an update to the earlier Trustworthy Repositories Audit & Certification (TRAC) [6.6]. This provides a series of criteria to show that the archival system can sustain information in the long-term. For example, TRAC criterion A1.2 is:

> *A1.2 Repository has an appropriate, formal succession plan, contingency plans, and/or escrow arrangements in place in case the repository ceases to operate or the governing or funding institution substantially changes its scope.*
>
> *Evidence: Succession plans; escrow plans; explicit and specific statement documenting the intent to ensure continuity of the repository; and the steps taken and to be taken to ensure continuity; formal documents describing exit strategies and contingency plans; depositor agreement [6.7]*

(Escrow arrangements typically involve a third party having a copy of the data, but only giving access to it if the repository fails.)

The criteria of TDR and TRAC are designed to give assurances that an external repository is a trustworthy place to store data. They can also be used for self-audit, to ensure that the system put in place actually meets the requirements of the business for long-term data sustainment.

6.6 **Summary**

The problem for governance is not merely that the executive board issues a fiat "preserve information for the long term," but ensuring the detail of that command meets the needs of the business and the resources to enact it are available. That requires several types of knowledge including business plans, the engineering use cases—describing what needs to be preserved for which designated community—and technical expertise to identify the most appropriate storage format. Plans for information preservation have to be costed. The implementation must be audited, not only to see if it is actually preserving what it is supposed to preserve, but also to see if its planning is adequate to meet the requirements in future. And part of that adequacy is ensuring that the provenance of information is maintained, since, in an aerospace context, this means demonstrating that the information has been generated through the correct processes to ensure that the aircraft is airworthy.

References

6.1   Wikipedia, "Enron," https://en.wikipedia.org/wiki/Enron, accessed May 15, 2019.

6.2   INCOSE, "Guide to the Systems Engineering Body of Knowledge," https://www.sebokwiki.org/wiki/Guide_to_the_Systems_Engineering_Body_of_Knowledge_(SEBoK), accessed May 10, 2019.

6.3   DeMarco, T. and Lister, T., *Peopleware: Productive Teams* New York: Dorset House, 1987 (further editions 1999 and 2013).

6.4   Addis, M., "Cost Models and Cost Modelling Tools," Retention and Access Services in Supply Chains (Project 100939), University of Southampton, 2012.

6.5   Centre for Research Libraries, "ISO 16363/TDR," https://www.crl.edu/archiving-preservation/digital-archives/metrics-assessing-and-certifying/iso16363, accessed May 14, 2019.

6.6   Wikipedia, "Trustworthy Repositories Audit & Certification," https://en.wikipedia.org/wiki/Trustworthy_Repositories_Audit_%26_Certification, accessed May 14, 2019.

6.7   Centre for Research Libraries, "TRAC Metrics," https://www.crl.edu/archiving-preservation/digital-archives/metrics-assessing-and-certifying/trac, accessed May 14, 2019.

Basics of CAD

7.1 What Is CAD?

If you work with an aerospace design office, it is clear that CAD plays an important part in the design of an aircraft. But what exactly is CAD? how is it used? and why does that make it any more difficult to archive than documents? This chapter explores CAD in just enough detail to explain why archiving is nontrivial, so that the next chapter can go into the detail of archiving it.

Computer-aided design (CAD) is a generic name for a range of tools that engineers use to develop designs. There are specialist applications for electrical design and manufacture, for developing mechanisms, for systems engineering, and so on. However, used without an explicit context, CAD means "mechanical CAD"—a system which mechanical engineers use to design the shape of a product's components—anything from a simple bracket through to a whole wing.

CAD was conceived in the 1960s as a form of electronic pencil.[1] However, unlike a piece of paper, which simply recorded where a pencil passed, the lines in a CAD system are recalculated each time the design is displayed. In terms of a drawing app, this is the difference between a freehand curve which follows the position of a mouse, and a straight line drawn by clicking at its two end points—and the more sophisticated apps allow you to pick up an end point and redraw the line elsewhere. Obviously, CAD goes beyond a simple line into 3D shapes, but the model can still be tweaked by altering its control points.

Modern design processes are based on "model-as-master," in which the 3D model is the authoritative statement of design, rather than a drawing derived from it. This

---

[1] The late John McDonnell of BAE Systems remembered the conference in the 1960s when CAD was first proposed, "What was needed was an equivalent of a draughtsman's spline."

model is used directly for analysis—for example, checking that a part can withstand the stresses it will be put under. And it is used for manufacture, the 3D shape being input to a computer numerically controlled (CNC) machine tool, which can carve out the part from a block of solid metal—or perhaps print it directly in 3D. Drawings and images of the model are not usually stored, but rendered directly from the CAD model.

The model is not simply a record of what was designed, its main purpose is to *communicate* the design. For example, it is sent downstream to manufacturing, who need not only the shape of the part, but also manufacturing information such as surface finish and the accuracy of the dimensions (*machining tolerances*). It will also be sent across time—the subject of this book—so that the design can be reassessed in the light of new information, or possibly reused.

This chapter dips into some of the technicalities that can stop models being transferred between different CAD systems and could therefore stop a CAD file being revisited after 20 or 40 years. Different types of curves show the importance of CAD systems having the right algorithms. Product & Manufacturing Information (PMI) is added to communicate to manufacture. Information standards such as Standard for the Exchange of Product data (STEP) explain precisely how a CAD model can be read into a different CAD system or a CNC machine, and how CAD systems with different software can use the same geometric knowledge.

7.2 Curves: From Data to Equations

The designs for aircraft, cars, and many other products require smooth curves for reasons ranging from the appearance of the product to avoiding stress or for achieving a smooth airflow across a surface. Surfaces are constrained by the needs of the volumes they contain—to cover an engine, to house passengers, and so on. The surface curves have to be easily calculable—if the equations are too complex, the arithmetic approximations inherent in computers mount up, distorting the curve.

The usual approach is to set out a series of control points which the curves must pass through (P1–P5 in Figure 7.1), then for each pair of control points, calculate a segment of curve. These segments must meet at the control points, and there must be no sharp change in direction there. This latter criterion is often reformulated as *tangent continuity*. Given these constraints we can briefly explore two of the simpler types of curves which might fit the bill: circular arcs and bicubic splines.

A circular arc has a simple formula:

$$x = x1 + R\cos(t), y = y1 + R\sin(t), t1 < t < t2$$

where (x1, y1) is the center of the circle, R the radius, t the angle, and t1 and t2 define the end points of the arc. To fit a circular arc between two points one also needs the tangent at one of the end points. This is exactly enough information to define the parametric equations above. Less information, and the center or the radius is undefined; more information and equations are over-constrained and some of the information has to be ignored.

However, since tangent can be set at only one end of the arc, therefore there will be tangent discontinuity where it meets the next arc. This problem is solved by biarcs, where each segment of curve is made up from two separate circular arcs. The two control points P1 and P2 are the base of the controlling triangle, and the tangent directions at each end define its other two sides (Figure 7.2). If the two arcs meet at the center of the controlling triangle, the arcs will also be tangent continuous [7.1]. The tangents at each

FIGURE 7.1 Defining a curve.

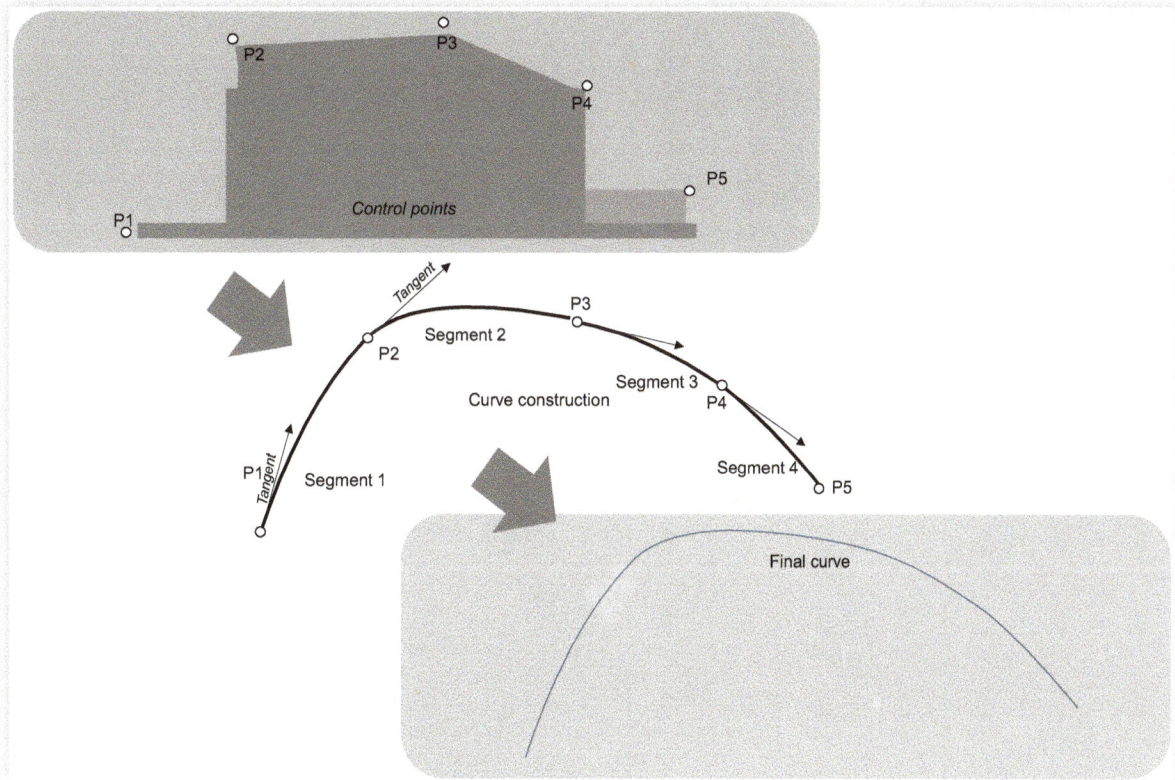

end are extended to form sides for the controlling triangles for the adjoining segments, so ensuring tangent continuity between segments.

However, although the surface is smooth, there is a step change in the radius R at each junction between each arc, and this is a curvature discontinuity (curvature is the reciprocal of radius). If you bend a sheet of metal to meet the tangents at the two ends of a segment, both the tangent and the curvature change smoothly. This smooth change of curvature is what the old, physical splines achieved. So, although biarcs are smooth curves, and useful for, say, designing the cutting patterns for shoes, they do not represent the curves created when bending metal to form the skin on an aircraft.

A different type of curve—the bicubic spline—gives the same curvature continuity as a physical spline. It is bicubic because the equations for each of the x and y components are given by cubic equations in the parameter t:

$$x = a_1 . t^3 + a_2 . t^2 + a_3 . t + a_4$$

$$y = b_1 . t^3 + b_2 . t^2 + b_3 . t + b_4 \quad 0 <= t <= 1$$

The spline also comes with a control polygon, which makes it easy to understand how to adjust the shape of the curve. The control polygon (Figure 7.3) is made up from the two control points A and D at the end of a segment, plus the two tangent lines at each end of the segment. However, unlike the biarc triangle, the lengths of the tangent lines are also important, and these have end points B and C. Lengthening or shortening

FIGURE 7.2 Constructing a biarc.

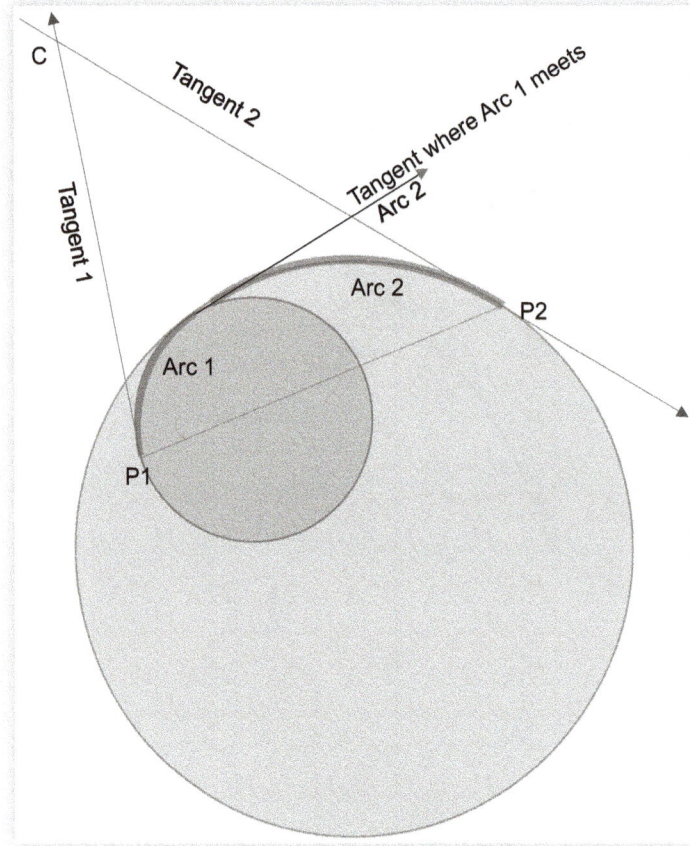

FIGURE 7.3 Control polygons for a bicubic spline.

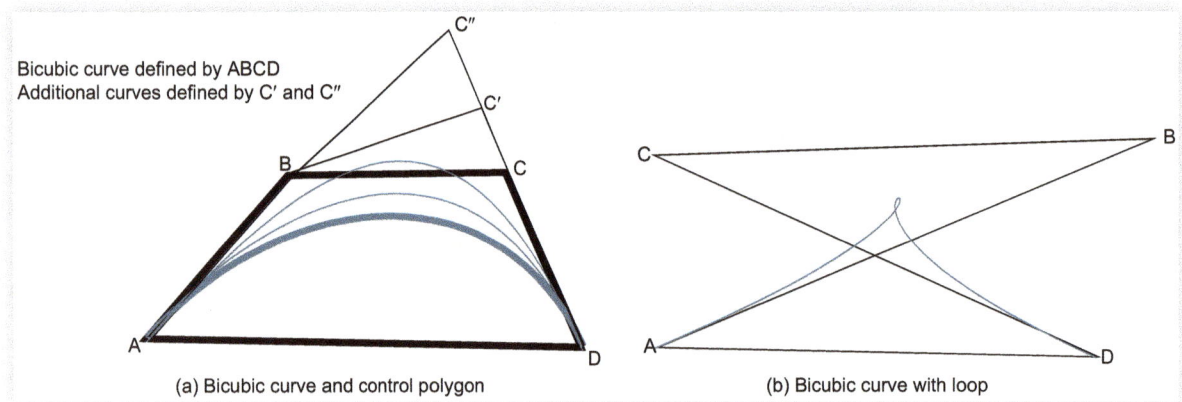

Bicubic curve defined by ABCD
Additional curves defined by C′ and C″

(a) Bicubic curve and control polygon

(b) Bicubic curve with loop

the tangents changes the shape of the curve. Making the tangent too long or too short can generate problems such as loops (Figure 7.3b).[2]

Does the average user care which formulation is used? No, not if they are simply designing for their own pleasure. But, Yes if they need to communicate to someone else. The CAD file contains only the constants that define the curve—x1, y1, R, t1, and t2 for an arc, or the values of the a's and b's for a spline—but the formulae used to calculate the curves are part of the CAD software. Feed the constants of one type of curve into a CAD package that uses a different type of curve, and, if it does not reject the file outright, the CAD package will produce a curve of a different shape and communication has not been achieved.[3]

In 3D, this grown-up version of a dot-to-dot puzzle can also be solved in several different ways, so communicating CAD designs relies on the sender and the receiver solving it in the same way, whether they are separated by an ocean or by a gap of 20 years. Information standards describe unambiguously how to specify each type of curve or surface. Currently, in standards, most curves are represented as non-uniform rational B splines, a generalization of the bicubic spline, but which can also represent the conic curves such as circles [7.4].

7.3 Geometry, Topology, and Errors in the Model

Figure 7.4a shows a badly made box. Six curved surfaces have been brought together, but geometry of the box is not right. The curves start and end in the same places—the corners of the box—but when the intermediate points are calculated, gaps appear. Such errors can arise because of the way that computers represent numbers. Even if computers work to 16 decimal places, the errors in complex calculations can quickly reduce the real accuracy to 4 places. The error on the result is the computational tolerance. This is distinct from the manufacturing tolerance, which is a function of how accurately the part must be made.

Figure 7.4b is a topological view for three of the six surfaces, showing how each surface connects to the others along its edges. Topological calculations are exact and have results that are true or false (does Surface 1 connect to Surface 2?), or have a whole number answer (how many other surfaces does Surface 1 connect to?). In theory, the topology of a 3D shape and its geometry should correspond, and if they don't there is something wrong. For example, if only half of Surface 2 was used (Surface 2′ in Figure 7.4c), then Surface 1 would not join to it, and the box would not be a box at all. CAD models check for such obvious errors. But, returning to the original box, is the gap between Surface 1 and Surface 2 an error by the modeler, or is it just the result of computer arithmetic? One way of resolving this is to put a tube around the two "connecting" edges (Figure 7.4d). If both edges fit within a tube that is smaller than the computational tolerance, then the solid is valid, otherwise the CAD system flags an error.

Historically, computational tolerances have thrown up errors when transferring a CAD model between different CAD packages, for example, if the gaps tolerated by the

---

[2] For a more detailed discussion see, for example, [7.2]

[3] This was a notable problem *for blend* surfaces used to smooth over the 3D joint between multiple intersecting surfaces. Some CAD software used proprietary formulae, making the usual transfer mechanisms impossible. Finding methods to transferring such surfaces was the subject of the Djinn project [7.3].

FIGURE 7.4   Geometry, computational error, and topology.

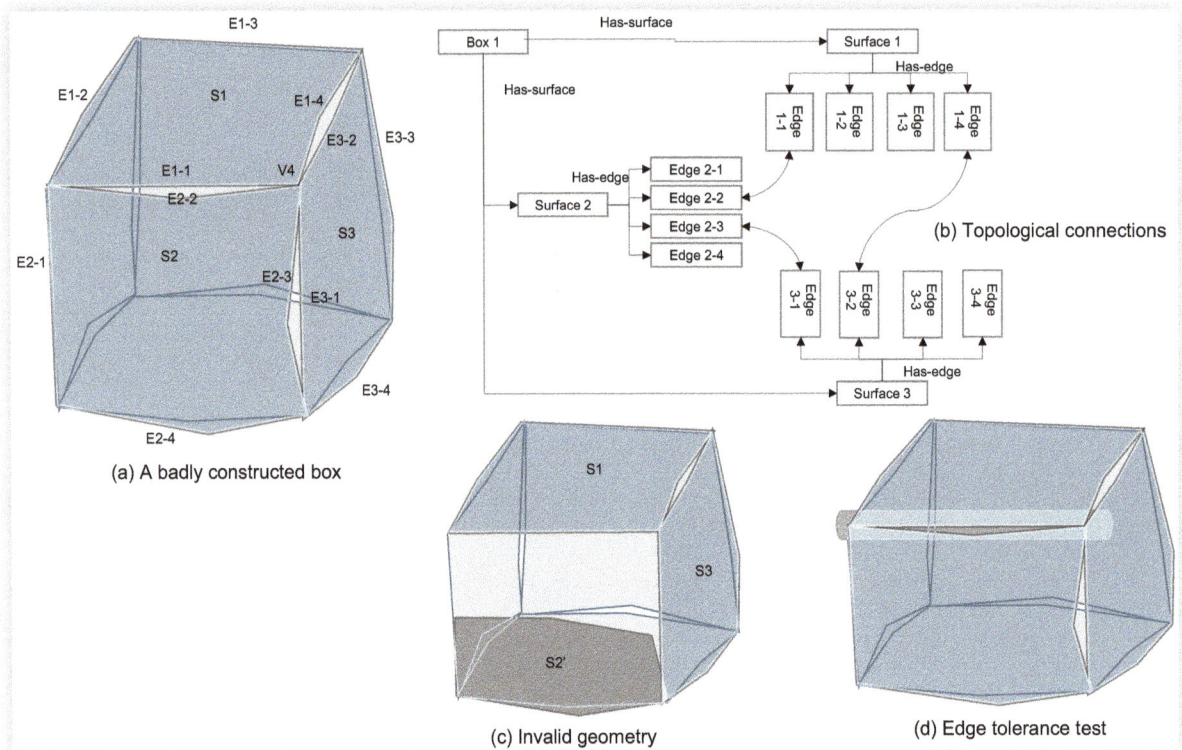

(a) A badly constructed box

(b) Topological connections

(c) Invalid geometry

(d) Edge tolerance test

first package are greater than the second package will accept. In long-term archiving, it is hard to go back in time and get the original model updated so that the current generation of software will accept it. Consequently, quality criteria should identify what computational tolerances are acceptable in the original model, and how the current CAD system should deal with models that do not meet its more stringent quality criteria.

7.4 Assemblies: More Than Just Parts

Most mechanical systems are not single parts, but assemblies of many parts. The individual parts are modeled separately, but to see what the whole assembly looks like—and ensure it fits together—it is necessary to bring all the separate parts together. There are three basic options for storing assembly models (Figure 7.5).

First, the assembly model may include the whole assembly tree (Figure 7.5a), and the stored model will hold the geometry of the individual parts and the positioning information for each part. This can limit the number of parts the model can contain. Second, the geometry of the individual parts can be stored in separate files, so the assembly model will store positioning information for each part but only reference the parts (Figure 7.5b). Third, the whole assembly tree can be broken into a series of levels, with an individual assembly model storing only the parts (or part references) and positioning information for one level, with higher-level assemblies being created to bring the lower-level assemblies together (Figure 7.5c).

The way assembly models are held will depend on the needs and practice of the organization creating the models, and archiving standards recognize this. Behind this there

FIGURE 7.5 Storing assemblies.

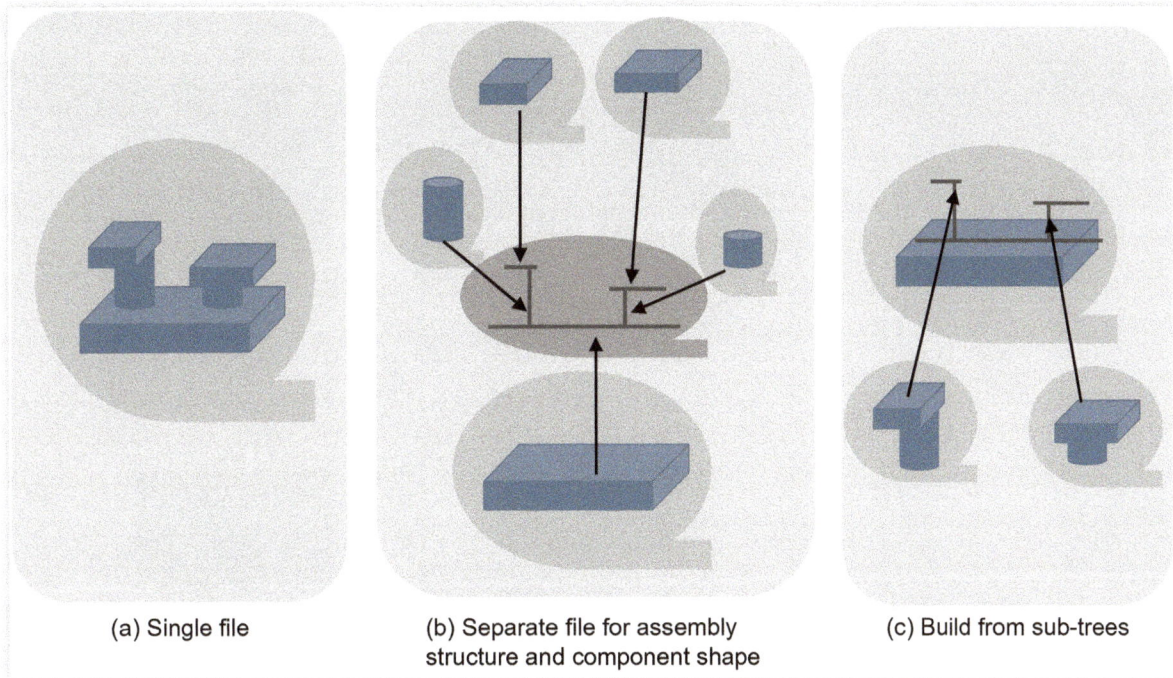

(a) Single file   (b) Separate file for assembly   (c) Build from sub-trees
structure and component shape

is a whole complex of procedures and modeling standards on how coordinate systems are set up and managed. These practices will need recording in the archive as part of the supporting information for the models and referencing in the archival metadata.

7.5 Types of CAD System and Their Incompatibilities

Not only are there different types of curve, but there are also different types of CAD systems, dividing roughly into *wire frame*, *surface modelers*, and *solid modelers*. A wire frame system represents only the lines of the design—the edges of a surface—as a draftsman would in a drawing. It leaves visualizing the surface to the imagination. If you were to cut a section through a wire frame model, you would get only a pattern of dots where you have cut through the wires (Figure 7.6a), not a profile of the surface.

A surface modeler represents the surfaces, not just the edges and the contours, although it does not represent the interior of any solid. Cut a model in half, and you get its profile at the cut (Figure 7.6b). One can calculate whether a point is inside or outside a solid by tracing a path from the space outside the model and keeping track each time it passes through a surface. Because the surfaces are the boundaries of a solid, it is also called a Boundary Representation or Brep. One variation in surface modeling is a surface modeler with topology, in which a topological model links each surface to the solid they bound, showing, for example, that six squares don't touch by accident but make up the surface of a solid cube. This makes it easier to verify that the different surfaces fit together with no gaps—for example, that the designer has remembered to include all six faces of a cube.

FIGURE 7.6   Types of CAD model.

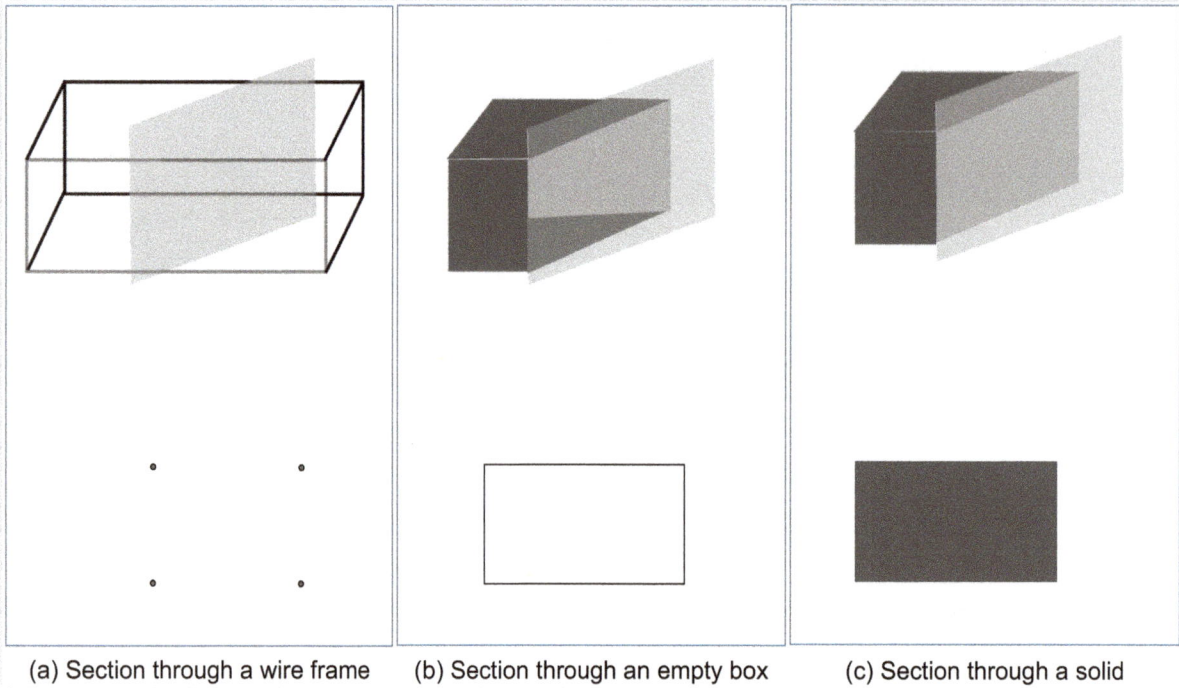

(a) Section through a wire frame   (b) Section through an empty box   (c) Section through a solid

In a solid modeler, the inside of a model appears solid (Figure 7.6c), and there is no need to track to the outside, and no possibility of leaving out one of the surfaces. Constructive solid geometry (CSG) [7.5] is the best-known type of solid modeler.

Moving from wire frame to surface to surface-with-topology and onto solid increases the information available within the model, making it easier to verify that the model is correct. Potential quality failures include gaps where surfaces should meet, faces on models which are too small to compute accurately, or having two identical solids in the same place. It is also much easier to reduce the information in a model—to convert from a solid model to a surface model or a wire frame—than to go in the opposite direction and try to infer missing information. However, early solid modelers could only represent a limited range of surfaces, whereas surface modelers had considerably more flexibility. Consequently, when the STEP standards were initially developed, they did not cover CSG modelers.

7.6 Manufacturing and the Information It Needs

Imagine you want to knock up some wooden shelves for the garage. A quick sketch (Figure 7.7a) will show how much wood you need, although when you get to the DIY shop you notice that the sketch does not show how deep the shelves are—and does the sketch show the width of the individual shelves or the overall width including the uprights? Then when you make the shelves, it does not matter if the end of the shelves are not quite square, or one is a couple of millimeters shorter than the others—they are only in the garage, not the front room.

FIGURE 7.7 Product manufacturing information: (a) rough sketch and (b) with PMI.

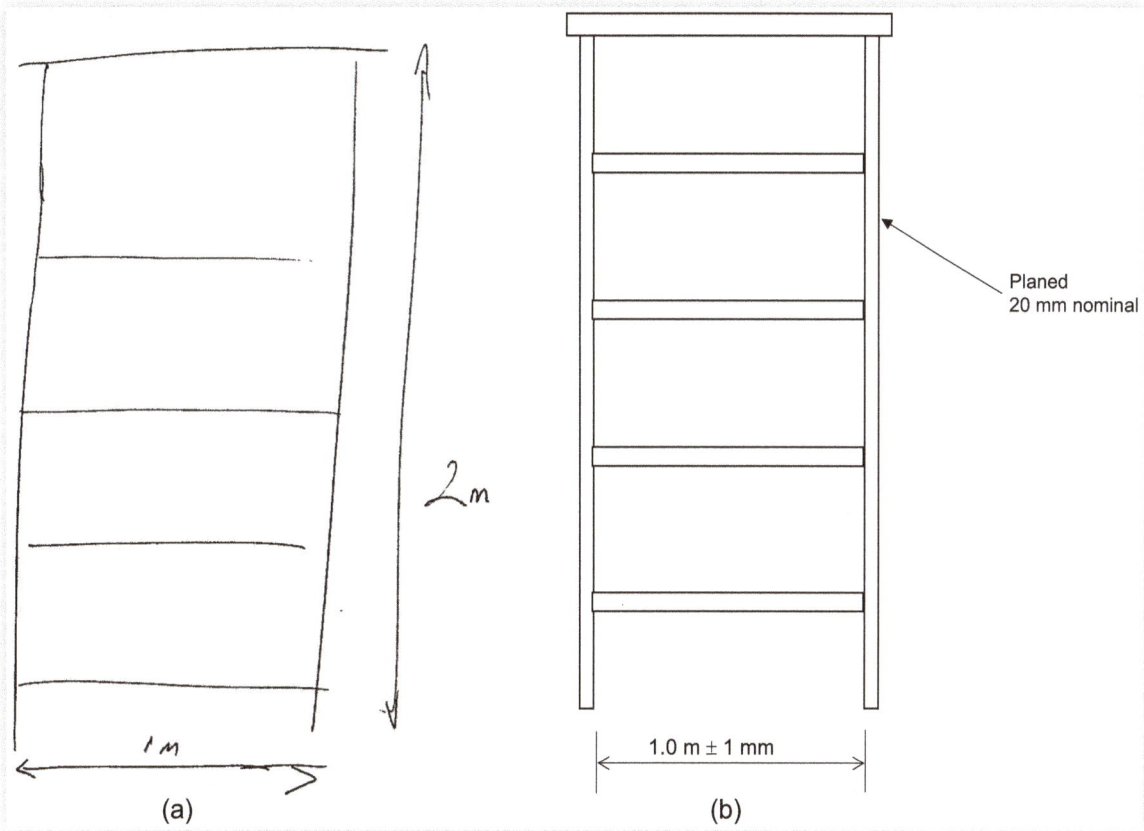

(a)   (b)

Planed
20 mm nominal

1.0 m ± 1 mm

Not surprisingly, when building an aircraft, these sorts of questions need more precise answers, and these are given through *Dimensions and Tolerances*, which are the core topics of *PMI*.

In a conventional drawing, the dimensions are shown by adding dimension lines to pick out what is being dimensioned (Figure 7.7b) plus an annotation giving the size (e.g., 1 m), and this can be accompanied by a tolerance such as ±1 mm—that is, the part should be between 0.999 and 1.001 m long. Tolerance figures come from the engineering knowledge of the designers: the function of the part, the effect of small deviations in manufacture, and the costs of both manufacture and assembly—it costs more to make a part with a smaller tolerance, but it is more certain to fit exactly.

In older CAD-model-for-drawings processes, the PMI information was simply additional lines and text overlaid on the shape model. In a modern CAD model, the user can interrogate the model to measure the size of any particular dimension. Annotations such as tolerances can be linked to a feature, enabling the user to pick out the annotation relevant to the feature they are working on. Some CAD systems will even include advanced algorithms to ensure the combined tolerances of the individual parts meet the required tolerance for the whole assembly. Moreover, in digital manufacture, the CNC program can be automatically generated from the design without the need for a human to read the dimensions or tolerances from a drawing.

PMI is a technical communication from a designer to the manufacturing department. To avoid ambiguity and ensure consistency, it is highly formalized [7.6], and set out in standards such as that from ASME [7.7]. There have been substantial developments in these standards in recent years, and the information standards for exchanging CAD have had to be updated to follow these developments.

7.7 **Visualization: CAD Lite**

CAD systems are primarily aimed at the design of mechanical systems. Aerospace companies invest very considerably in training designers to use these tools. Visualization tools provide a relatively lightweight, low-cost facility to view CAD files and are less expensive and need much less training.

Visualization tools typically use an approximation to the exact shape, such as a faceted polygon in which the original smoothly curved surface is approximated by small flat faces, often triangles or rectangles. Because of the importance of the gaming market, computer graphics chips are designed to rapidly color-shade each facet in a way that makes the surface appear smooth, and these can greatly speed up the rendering of faceted CAD models. "Lightweight" refers to the way that the resulting files are smaller—often only a few percent of the original CAD file—and require much less computing power to render. This allows relatively large and complex CAD models to be displayed on relatively low-cost hardware.

Visualization tools do not generally give the ability to create or modify the shapes, but do allow 3D operations such as rotation. Frequently they will also provide CAD-like functionality, such as turning on/off annotations, or picking out a single part from an assembly. They may also allow the user to "measure" the model, providing dimensions, which may not be explicitly called out in PMI. Consequently, they can be used in processes downstream of design, such as manufacture or the documentation of maintenance procedures.

A variety of visualization format have been developed. JT Open was originally a proprietary format, but which is now an ISO standard [7.8] and is being promoted as a format for long-term archiving, although there are still concerns about whether it supports the requirements of an open standard. Some .pdf file formats allow the inclusion of 3D images, in which CAD files can be included in documents, for example, a maintenance manual. A number of XML formats have also been developed.

7.8 **Defining CAD Data: Exchange Standardization**

The aim of CAD data exchange is to take a CAD file produced in one company using their preferred CAD software and send it to another company—for example, a supplier—who can then read it into their preferred CAD software. A standardized CAD format allows a CAD file to be transferred to any other package supporting the standard—including those not written yet. In aerospace, as with several other industries, the key standard is STEP.

The STEP is a complex standard issued in many parts as ISO 10303. The initial parts focus on how the standard describes data and includes the data modeling language EXPRESS (ISO 10303-11), a text file format (part 21), and an XML binding (how the data is translated into XML). It also contains a number *of Integrated Generic Resources* and *Application Interpreted Constructs* (AIC), which provide common definitions for many of the key geometric constructs.[4]

Data exchange is through an *Application Protocol* (AP), in which an Application Activity Model (AAM) identifies the process for which the AP is designed, the Application Reference Model (ARM) maps from the process to a data model, and the Application Interpreted Model (AIM) provides the detailed model for implementation. The AAM is important as it provides the context of data definitions, since, for example, "part number" can have different interpretations as one moves from design to manufacture and into support. The ARM is important for tying down the business concepts that are embedded as contextual information within the exchange file, for example, "approval" and "approval status." Over recent years, particularly with more business process-oriented standards such as Product Life Cycle Support (Part 239), the ARM has become more important, and may be the target of implementation. However, for geometry the AIM has remained key.

Historically, the most important Application Protocol from the aerospace perspective has been AP 203, "*Configuration controlled 3D designs of mechanical parts and assemblies.*" Its key differentiator over earlier standards was the inclusion of conformance classes. This made mandatory the support of all the data entities in any particular conformance class, and AP 203 had six classes:

1. Configuration Controlled Design Without Shape
2. Class 1 plus Geometrically Bounded Wireframe and Surface Models
3. Class 1 plus Wireframe with Topology
4. Class 1 plus Manifold Surfaces with Topology
5. Class 1 plus Faceted Brep
6. Class 1 plus Brep

There are three things to note here. First, that conformance class 1 allowed for the exchange of process data ("metadata") used to manage the development of a CAD model. This would later form the basis for the STEP PDM Schema (see Chapter 10). Second, conformance classes 2–6 dealt with the different types of CAD model (see Section 6.4), which therefore identified to users whether they could exchange their CAD model with another package. And third, when the standard was developed, solid modelers were not mainstream and were not covered by the first edition of AP 203.

In the aerospace environment, the initial tests of AP 203 fell foul of different quality criteria within the CAD modelers themselves. One of the main CAD packages of the period was still using single precision arithmetic, while most others used double precision, and they detected "gaps" between the surface patches and rejected them. This was the first of several practical lessons learned about how differences in software could lead to differences in model geometry. Three strands of development followed from this.

First, the implementer forum CAx-IF was set up to test CAD-to-CAD file exchange [7.13]. Tests are done by CAD package developers and by developers of translators, and reported back to the supporting organizations, including PDES Inc. and ProSTEP iVIP (see note 5.1), who then report back to their member companies. Test rounds cover the continued compatibility of new versions of translators and initial testing of new STEP features.

Second, CAD quality tools were developed, sometimes as separate software suites. These detect both poor practices by the designers and problems that arise through translation. Designer errors range from putting the same hole in the same place twice,

---

[4] Reference [7.9] provides a convenient starting point for details on STEP. See also [7.10, 7.11, 7.12].

to forgetting to save the model at all. Numerical problems include gaps that appear in the surface after translation, and here the translator software may be extended to provide automatic surface healing.

Third, the first *validation properties* were developed, namely the volume and the surface area of a model. These properties are generated in the originating CAD system, added to the STEP exchange file, and then regenerated in the receiving CAD system using its own algorithms. If the before and after properties differ significantly, this is a strong indicator that something has gone wrong with the exchange. One LOTAR partner also reported informally that major validation errors were strongly correlated with poor practices by designers.

Although AP 203 is extensively used through the aerospace sector, the European auto-motive sector has preferred AP 214. Both use the common geometrical constructs from STEP. AP 203 and AP 214 have now been brought together in AP 242 (see next section).

The biggest problem for the STEP APs is that they work really well and so the STEP interface gets taken for granted. If some aspects of CAD model exchange don't work, this is seen as a failure of the standard, whereas usually the problem is that the users failed to engage with the STEP development process, and have not got the features they want into the standard.

7.9 **Standardizing CAD Data: STEP AP 242**

AP 242 "Managed model-based 3D engineering" was kicked off by users in the LOTAR project because they realized that they wanted more functionality in STEP than they could get from existing APs. It takes forward both AP 203 and AP 214 (and it is no coincidence that 42 = 3*14) ([7.14], see also [7.15. 7.16, 7.17, 7.18]). As well as the geometry and model management found in the original APs, it includes:

- **presentational and representational PMI**,

- **tessellation**, allowing STEP to support a light visualization model in an efficient way,

- **composite structures**, **electrical harness**, and **piping**,

- **product data quality** (quality criteria and check results),

- **mechatronics**

Presentation versus representation is a key distinction in STEP. A presentation is a rendering or view of information as a human readable image. This means that if software needs to access the value of, say, a dimension or a tolerance found in presentational PMI, it must either ask the user to read the drawing and type the value in, or use optical char-acter recognition on the drawing. By contrast, a representation is intrinsically machine processable. This makes it easier to validate automatically, and allows a design to be fed directly into a CNC machine for manufacture.

The need to accommodate various organizational viewpoints, and also to keep key players such as the CAD vendors in the loop, makes standards development a relatively slow process. To get functionality available faster, AP 242 has been developed incre-mentally. The first edition was published as a full ISO standard in 2014, and the second version planned for late 2019 [7.19], which will include additional design models such as additive manufacture and electrical harnesses.

7.10 **Summary**

One of the key questions for long-term data sustainment is how knowledge is maintained over the passage of time, in order that the records of passed times may be read and understood. The Rosetta Stone had records in Egyptian hieroglyphics and Coptic Greek, but, because scholars had only been interested in Greek, only the Greek could be read—at least until it was worked out how to use the Greek to translate the hieroglyphics. The story of the Rosetta Stone illustrates how data (hieroglyphics) can be lost because the knowledge to read it is lost, as well as how it is retained (Greek) because some knowledge flowed down the ages.

The flow of the CAD model is a data flow, going from design to analysis, manufacture, and support. CAD sustainment requires also that the model flows across time to designers, regulators, and lawyers (see Chapter 2). However, these future communities also need the knowledge of how to read and interpret the CAD model (Figure 7.8).

This knowledge flow covers both the geometric procedures and the starting data for each procedure. For a bicubic curve, procedures are realized as equations in software, and the starting data as the bicubic coefficients. This starting data is formalized in information standards such as STEP AP 242. The information standards provide a common template, which software developers use to map geometric data into the starting data for their implementation of the geometric procedures. Obviously, the people who write the standards also know their geometry.

Data is the physical realization of a CAD model and provides the physical flow of the CAD model. Knowledge flows from the geometers—from Euclid onward—to CAD software writers and standards developers. Information is a mapping from a knowledge (geometry) to data (the CAD file), which binds the two flows together.

FIGURE 7.8 Knowledge flow in CAD.

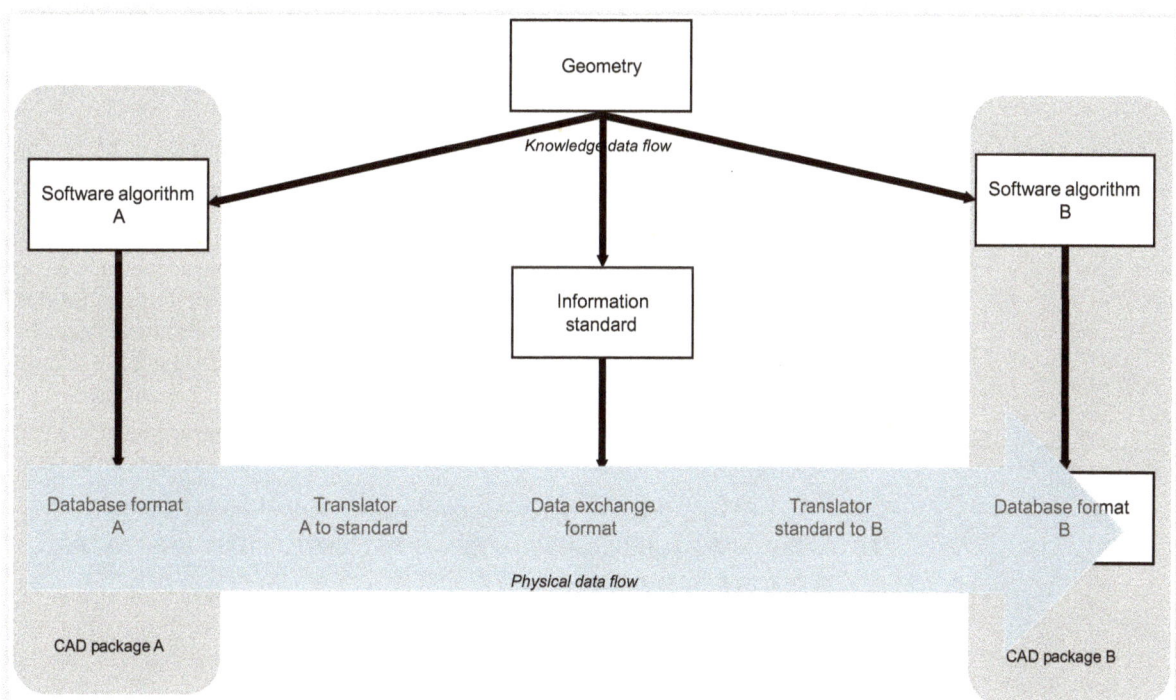

References

7.1 Sabin, M., "The Use of Piecewise Forms for the Numerical Representation of Shape," Reports 60/1977, Computer and Automation Institute Hungarian Academy of Sciences, 1976.

7.2 Wikipedia, "Spline (Mathematics)," https://en.wikipedia.org/wiki/Spline_(mathematics), accessed April 18, 2019.

7.3 Armstong, C., Bowyer, A., Cameron, S., Jared, G. et al., *Djinn: A Geometric Interface for Solid Modelling* (Winchester: Information Geometers Ltd, 2000), ISBN: 1-874728-13-5.

7.4 Wikipedia, "NURB," https://en.wikipedia.org/wiki/Non-uniform_rational_B-spline.

7.5 Wikipedia, "Constructive Solid Geometry," https://en.wikipedia.org/wiki/Constructive_solid_geometry, accessed April 12, 2019.

7.6 Shrinivasan, V., *Theory of Dimensioning* (New York: Marcel Dekker Inc., 2004), ISBN: 0-8247-4624-4.

7.7 The American Society of Mechanical Engineers, "Dimensioning and Tolerancing," Y14.5, 2018, https://www.asme.org/codes-standards/find-codes-standards/y14-5-dimensioning-tolerancing, accessed July 19, 2019; see also Wikipedia, "Geometric Dimensioning and Tolerancing," https://en.wikipedia.org/wiki/Geometric_dimensioning_and_tolerancing, accessed July 19, 2019.

7.8 Wikipedia, "JT (visualization format)," https://en.wikipedia.org/wiki/JT_(visualization_format), accessed April 11, 2019; ISO/DIS 14306:2017, "Industrial Automation Systems and Integration -- JT File Format Specification for 3D Visualization."

7.9 Wikipedia, "ISO 10303," https://en.wikipedia.org/wiki/ISO_10303, accessed April 11, 2019.

7.10 STEP Tools Inc., "The STEP Standard," https://www.steptools.com/stds/step/, accessed April 11, 2019.

7.11 IDA STEP, "About STEP (ISO 10303) Standard," https://www.ida-step.net/support/resources/about-step, accessed April 11, 2019.

7.12 Pratt, M.J., "Introduction to ISO 10303—The STEP Standard for Product Data Exchange," NIST Technical Note, https://ws680.nist.gov/publication/get_pdf.cfm?pub_id=821600, accessed April 11, 2019.

7.13 CAX-Implementor Forum, https://www.cax-if.org/, accessed April 12, 2019.

7.14 ISO 10303-242:2014, "Industrial Automation Systems and Integration — Product Data Representation and Exchange — Part 242: Application Protocol: Managed Model-Based 3D Engineering," https://www.iso.org/standard/57620.html, accessed April 16, 2019.

7.15 "AP242 Benchmarks," http://www.asd-ssg.org/step-ap242-benchmark, accessed April 16, 2019.

7.16 "STEP AP242," http://www.ap242.org/ Project, accessed April 16, 2019.

7.17 "STEP AP242 Benchmark #2 by AFNeT and ProSTEP iViP," http://benchmark.ap242.org/, accessed April 16, 2019.

7.18 http://www.ap242.org/composite-design-interoperability.

7.19 http://www.ap242.org/edition-2, accessed April 16, 2019.

8

Preserving CAD

8.1 What Is Your Goal?

You are a CAD manager somewhere in an aerospace supply chain. Your executives have decided to keep CAD for at least 40 years. The IT department have read OAIS and bought some new disk drives. The inevitable Web search has brought up a useful overview such as "Preserving Computer-Aided Design" [8.1, 8.2], while the budget is there to buy copies of the LOTAR standard. What more do you need? Actually, quite a lot, for which this chapter provides an overview.

LOTAR part 100 provides a general introduction to archiving of CAD. It covers the fundamentals of long-term archiving of 3D CAD and has a number of informative annexes, including "The Evolution of CAD Systems," "the main types of CAD," and "Considerations for long-term preservation of CAD" (essentially a quick guide to what can go wrong).

Later parts in the 100 series go into the specifics. They contain the condensed knowledge of some of the top experts who have been working in CAD for 20 years or more. While this means that they will probably have considered the sort of CAD model you need to archive, they have also considered many more. To create a standard, they have focused on the most common and general use cases for the sector, starting with explicit 3D geometry—pure geometric entities from points through to solids. This is described in part 110. Subsequent parts build on earlier parts, so that part 115, Explicit CAD Assembly Structure, adds assemblies to the guidance of part 110. Later parts go into detail for additional CAD use cases, such as the 120 series for PMI and the 130 series for machining form features.

Each specific part covers five main topics:

- Use cases defining the business context
- Essential information—what has to be preserved to meet the use case
- Core model—how that information is preserved
- Verification rules to check that what is saved is of sufficient quality
- Validation rules to check that when the model is read from the archive, it is unchanged

The parts also contain informative annexes, which provide additional background and explanation, for example Annex F of part 110 goes into detail about geometric validation properties.

However, you will still have to tailor the LOTAR guidance to your business need. For example, if very high integrity is required, then LOTAR recommends that you use an extended set of validation and verification criteria. This certainly means learning how to set the test thresholds for the quality criteria and probably means additional testing to show the values used meet the business need.

The remainder of this chapter goes into some of the detail of archiving for explicit geometry and assemblies.

8.2 Preserving Basic 3D Geometry

8.2.1 The Core Geometry

LOTAR part 110 covers the essential information for CAD 3D geometry [L 110 section 1]. It covers four scenarios [L 110 section 5] for each of the four basic use cases mentioned as requirements in Chapter 2 (certification, product liability, reuse, and product support):

- Exact geometry only
- Tessellated geometry only
- A part file containing exact and tessellated geometry
- A part file containing an assembly mixing exact and tessellated geometry

The essential information consisting of solids, surfaces, curves, and points. In more detail [L110 section 6], this is:

- The exact boundary representation of the shape of a single part
- The tessellated boundary representation of shape
- The exact and tessellated representation of surfaces
- The exact and tessellated representation of curves
- The representation of points

It excludes:

- Implicit or parametric geometry
- Geometric dimensioning and tolerancing and PMI
- Assemblies

- The presentation of explicit geometry (e.g., screenshots and 2D drawings)

- Machine readable rules for verification and verification

Implicit geometry is, in effect, a parameterized design procedure. If the parameters are updated, the model is recomputed [L100 Annex B]. For example, an extrusion is a procedure that takes a 2D profile and creates a 3D solid by moving it along a line—a cylinder can be made by extruding a circle. Preserving implicit geometry would require that the design procedure is archived and that any target CAD system can interpret that procedure. Parametric geometry captures the design intent in constraints represented by mathematical equations. The exclusion of these methods does not prevent any implicit or parametric geometry being converted to explicit geometry so the explicit form can be archived.

The core model [L110 7] for this geometry is detailed in STEP in ISO 10303-514 (advanced boundary representation) and ISO 10303-42 (geometric and topological representation). For practical purposes, the export from the CAD system will use one of the STEP Application protocols that use these resources:

- AP 203 conformance class 8 (CC08)

- AP 214 conformance class 2 (CC02)

- AP 242

The core model describes what *should* be archived, but, if the AP supports it, other information *may* be archived in the same file; for example, design history, layers, auxiliary geometry, or attributes such as surface roughness. However, these should be validated and verified using the rules in later LOTAR parts.

8.2.2 Verifying the Core Data

Verification tests whether the process of converting the CAD model into STEP creates a valid model, for example, one without gaps between faces. Verification is part of the ingest process and is applied to the STEP file. For geometry [L110 8], verification level 0 means no tests are done at the risk of failure when the model is retrieved. To reduce risk of retrieval failure, verification level 1 uses rules defined in SASIG PDQ [8.3] and ISO 10303-59 "Quality of product shape data." If the business needs to reduce risk further, additional rules can be applied (verification level 2). The LOTAR project tested the verification rules using a number of CAD tools [L 110 Annex D] and made the following recommendation (Tables 8.1 and 8.2):

TABLE 8.1 Level 1 verification tests.

SASIG code	Rule content	Comment/recommended threshold
G-CU-LG	Large Segment Gap	≥0.001 mm
G-CU-NT	Non-Tangent Segments	≥75 degrees
G-LO-LG	Large Edge Gap	≥1.0 mm
G-SH-FR	Free Edge	Not valid for a solid body
G-SU-NM	Over-used Edge	≥3 (count)
G-SU-FG	Fragmented Surface	≥2,500 patches
-	Empty Model	No solid or surface geometry

TABLE 8.2 Level 2 verification tests.

SASIG code	Rule content	Comment/recommended threshold
G-FA-EG	Large Edge Face Gap	≥1.0 mm
G-FA-TI	Tiny Face	<0.00001 mm² or area cannot be calculated
G-FA-VG	Large Vertex Face Gap	≥1.0 mm (not gap between vertex and edge)
G-SH-LG	Large Face Gap	≥1.0 mm
G-SO-EM	Embedded Solids	≥2 (count) with a proximity tolerance set to 1.0 mm
G-SU-TI	Tiny Surface Patch	<0.000 000 01 (only in surfaces with multiple patches)

The threshold levels were adjusted using the results of trials, based on whether there was a correlation between the threshold and test failing. These thresholds result from the computational tolerances of the CAD modeler software (see Section 7.3)

If the archive updates the stored form of the model or creates a different form for access, then the resulting model should be reverified.

8.2.3 Validating the Whole Model

Validation tests whether the model retrieved is the same as the model that was archived. It is a test of the CAD software reading the model rather than of the data and is there to ensure that when the CAD package reconstructs the full geometry, it does so correctly. Validation rules are based on properties that are computed from the geometry, such as volume, rather than properties that are represented in the geometry, such as faces being connected to each other. As with verification, LOTAR identifies three levels of validation from level 0 (no validation) to level 2, enhanced validation. It defines the following tests at level 1 [L110 9.3] (Table 8.3).

The properties themselves are described in the CAx-IF recommended practices [8.4]. These recommended practices also cover *Cloud of Points*, where additional points are placed on the surface and edges of the 3D solid in the original model and recorded with the model. On reading the model, the retrieving CAD package then checks that the points are still on the surface or edges—if not, the shape must have changed. *Cloud of Points* is a level 2 validation test, so it is up to the business whether to use it, and if it does, how to choose the points.

All the validation tests are subject to computational tolerances and variations in the algorithms, which calculate the properties. Pass/fail criteria depend on whether difference in the initial and final values is less than a threshold, but the threshold depends on various factors including the type of model and the application. For example, acceptable variations on a turbine blade will be less than for the internal structures for a cabin.

TABLE 8.3 Level 1 validation properties.

Geometry	Validation test
Exact solid	Volume, center of gravity, area
Exact independent surface/open shell	Center of gravity, area
Exact independent curve	Independent curve length
	Independent curve centroid
Tessellated solid	Surface area, centroid
Tessellated surface	Surface area, centroid
Tessellated curve	Total length, centroid

LOTAR [L100 7] puts the onus on the business to qualify-as-airworthy parts brought back from the archive and to set the thresholds according to their qualification needs. In fact, LOTAR requires qualification of the whole process for archiving as meeting aircraft quality standards. This in turn requires that the business archive their qualification rules and thresholds.

8.2.4 And What Else Do You Need to Remember?

Sending information to an OAIS archive requires the creation of a submission information package (SIP). LOTAR part 110 describes how the content information is stored using STEP as the archiving format, and the verification and validation tests needed to reduce the risk that the content becomes unreadable over time. However, the SIP is more than just the content information, and it must contain descriptive information to make the content information usable. Additional information is collected in the ingest process, and needs to be stored in the archive information package (AIP) (L100 10). This information includes

- Reference information

  - **part identifier***
  - **part version***
  - program or project identification*
  - model unique identification*
  - model issue (or design iteration)*

- Context information

  - **(STEP AP and Conformance class*****)**
  - **reference to the STEP standard—AIC 514 and AIC 512 for 3D geometry**
  - **reference to the LOTAR part used—here part 110***
  - **originating CAD system—name and release version**
  - STEP recommended practices—including CAX-IF guidance

- Provenance

  - **identification of the source file or model***
  - **conversion information—date, native format, and archived format**#*
  - **link to the report on the conversion** (see LOTAR part 11, 12)#*
  - **link to the verification report**#*
  - conversion process identification (name and version)#
  - process identifier for preparation and ingest (these will identify the implementation of the process in the creator's business)
  - identification of the person or organization creating the SIP
  - data validation report

- Fixity information
  - digital signature
  - encrypted checksum of content information
  - **validation properties**

- Other descriptive information
  - ATA chapter and zone

In the list above, items in bold are required, other items are optional. Items marked *
are also part of the descriptive information. Items marked # are created by the ingest
process. Details of the STEP AP and conformance class are embedded in the STEP file
and do not need to be recorded separately in the SIP but need to be extracted and added
to the descriptive information.

8.3 Preserving Assemblies

8.3.1 What Is Being Archived?

LOTAR part 115 covers the explicit representation of the way parts are brought together
in an assembly—that is, what parts are brought together in an assembly, and the 3D
positioning information that puts them in the right position relative to each other.

It does not cover the 3D shape of the part, as that has already been covered in part
110, although a single CAD file may contain both the geometry of the parts and the
assembly information. In this case, both parts 110 and 115 apply. Part 115 does not cover
"assembly-by-constraint," where, for example, the position of part Y is "perpendicular
to part X."

Within the business use case there are two main use cases (LOTAR's use of "use
case" is not always consistent). The first has a single submission for the whole assembly,
while the second has a number of incremental submissions, each containing a partial
assembly, starting with the lowest level assemblies and later submitting the higher-level
assemblies, with the top level being the last submitted. In the second use case, when
archiving a higher-level assembly, the submission process must check that the lower
level assemblies and the parts it references are already present within the archive. STEP
allows for top-down development through "promissory usage," but this is for use with
a product structure rather than a CAD assembly.

Within the use cases, LOTAR also allows for a number of variations, depending
on whether the assemblies are submitted as single or multiple files, and whether the
archival packages contain a single content file or multiple content files. Here its general
recommendation can be summarized as "keep things simple."

Figure 8.1 shows the essential information. Figure 8.1a shows the essential information
where the child node is held in a separate file. Figure 8.1b shows the case where a child
assembly is in the same file as the parent. This structure assumes that a single part is always
held in a separate file. "Occurrence identification" is relevant where the same part is used
several times in an assembly, and it is required to be able to distinguish each occurrence.

The essential information is represented in a core model based on the version 1.2 of
the STEP PDM Schema [8.5]. The PDM Schema harmonizes the interpretation of AP 203
and AP 214, and ensures data in the two schemas is interoperable. References to parts
and files that are outside the content file follow "Recommended Practices for External
References with References to the PDM Schema Usage Guide" [8.6].

FIGURE 8.1 Essential information for an assembly: (a) child in a separate file and (b) child in same file.

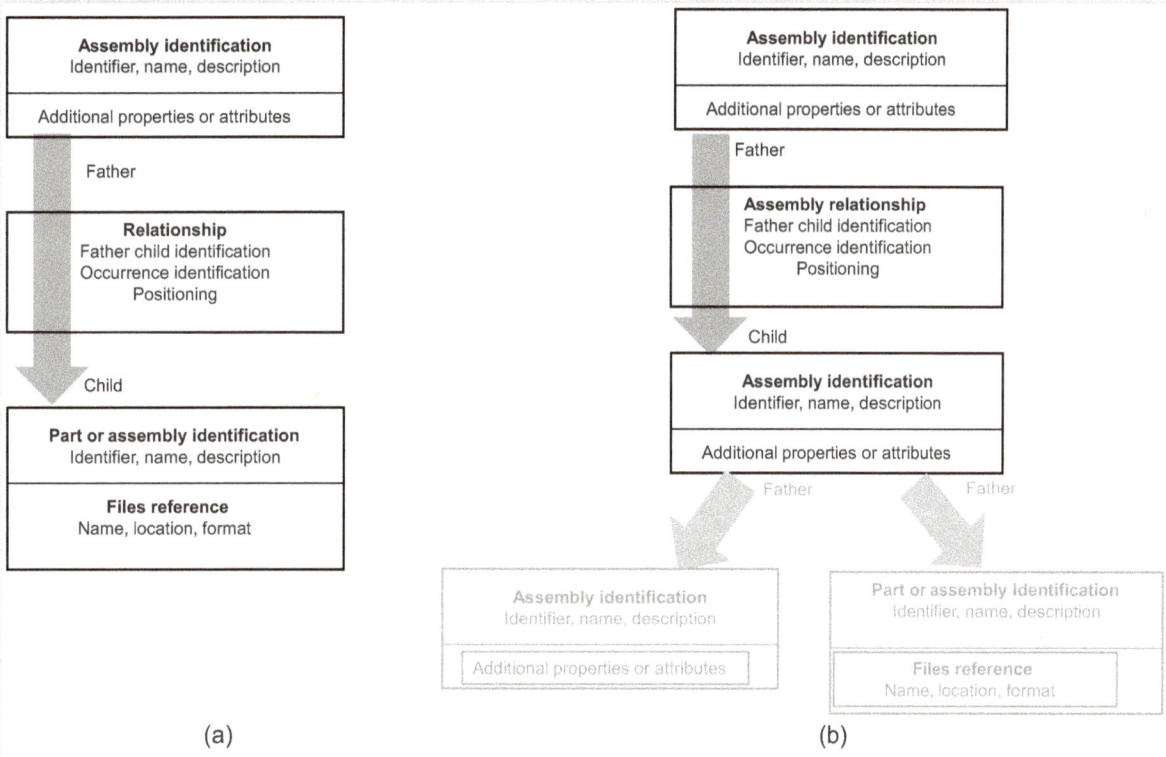

(a)   (b)

Note that the guidance is for an assembly tree, and not for a product structure. In particular, it does not include *effectivity*, which is used to manage design variants (see Chapter 10).

8.3.2 Verification Rules

For level 1 verification, the assembly structure must:

- Contain exactly one assembly

- There must be no orphan nodes not part of the assembly

- The assembly must be acyclic, meaning that the tree below any particular node may not contain a reference to that node, otherwise tracing down the tree will cycle round forever

- If the same assembly occurs multiple times, then it must contain the same parts, and in the same relative position (this is also a basic configuration control rule)

- Every part and assembly must have a unique identifier

LOTAR allows for additional rules for Level 2 verification but does not identify any.

8.3.3 Validation Rules

If the full geometry of the assembly is archived, then the geometric validation properties [8.4] of volume, area, and center of gravity are required to meet Validation Level 1. If the geometry is not archived with the assembly, or in the case of incremental archiving, the "number of children" and the notional centroid rules are used. The "number of children" is the sum of the number of occurrences of each child node recorded at each level of the assembly.

8.4 **Summary**

The domain-specific parts of LOTAR go into considerable detail on how to create model files that have a good chance of being readable in the long term. To do this, they advocate archiving data in a neutral format, and in the case of CAD, this is STEP. This is not to say that you shouldn't archive design models in the native format of the CAD tool that created them, but in the long term, the retrievability of those models will depend on the business model of the company that wrote the software, not the needs of your business.

Getting into detail, the domain-specific parts for explicit 3D CAD identify

- The use cases that the archive aims to support
- The essential information needed by those use cases (geometry, assembly structure)
- The core model that represents that information (using AP 203, AP 214, or AP 242)
- How to verify that the data recorded is correct
- How to validate the model as unchanged when it is retrieved

The verification and validation criteria have been piloted through the CAx-IF, which also provides guides on the criteria.

In addition, LOTAR part 100 provides the general context for the 100 series, including an outline of the descriptive information needed to find, interpret, and trust the archived files.

References

8.1   Alex, B., "Preserving Computer-Aided Design," Digital Curation Centre, 2013, https://www.dpconline.org/knowledge-base/80-cad.

8.2   Mike, A., "Untangling the Knot of CAD Preservation," Library of Congress, 2014, accessed July 16, 2019, https://blogs.loc.gov/thesignal/2014/08/untangling-the-knot-of-cad-preservation/.

8.3   ISO, "SASIG Product Data Quality Guidelines for the Global Automotive Industry," ISO/PAS 26183:2006.

8.4   CAX-IF, "Recommended Practices for Geometric and Assembly Validation Properties," Release 4.4, August 2016, accessed July 17, 2019, https://www.cax-if.org/documents/rec_prac_gvp_v44.pdf - Part 115 references geometric validation properties and assembly validation properties as separate documents, but they appear to have been combined.

8.5   CAX-IF, "Usage Guide for the STEP PDM Schema V1.2," January 3, 2002, available at the CAX-IF (confusing labeled with the release code V4.3), accessed July 18, 2019, https://www.cax-if.org/documents/pdmug_release4_3.pdf.

8.6   CAX-IF, "Recommended Practices for External References with References to the PDM Schema Usage Guide," Release 2.1, January 19, 2005, accessed July 18, 2019, https://www.cax-if.org/documents/rec_prac_ext_ref_v21.pdf.

Signposts for Other Models

9.1 Signposts: Not Answers

CAD provides a paradigm for archiving design models: complex data structures, but representable using standardized data models, supported by computable quality checks, with model integrity demonstrated through validation properties. Moreover, the LOTAR parts on CAD provides LOTAR's own paradigm for archiving: develop use cases, standardize the straightforward ones, and extend the standards to support the more complex ones.

This section briefly reviews some of the other types of design model considered for archiving. However, archiving and engineering are both large subjects and we cannot go into detail, rather we leave signposts that the reader may follow.

9.2 Requirements and Systems Engineering

Requirements and systems engineering are two disciplines which have their own archiving requirements, but there is coupling between them and an overlap of the standards that makes it convenient to deal with them together.

A requirement identifies what the customer wants. Engineering requirements are the result of taking the customer's commercial or functional aims and turning them into achievable performance figures [9.1]. These are what the design is costed against, and when, part way through development, it is proposed to change the requirements, they are the baseline for the additional costs. Consequently, although the engineering requirements are the starting point for design, a requirements model persists through the aircraft lifecycle for use in the management of change.

Requirements modeling can be summarized as a two-step process. First, general statements are broken down into "atomic" requirements—precise, unambiguous, single subject statements. Second, the atomic requirements are linked to the system that will fulfill those requirements. The linkage allows any change to the requirement to be traced through to the system it affects.

Historically aircraft systems were separately developed—fuel systems, power systems, control systems, and so on. Common methods and methodologies in each of these domains have been abstracted to form the foundation of *systems engineering*. Systems engineering is now mainly concerned with the functions and behavior of a system, rather than the physical or manufacturing design. This entails modeling multiple aspects of a system, including its structure, timings, and responses to triggers from controls and sensors, the dynamics of the system, and so on. Systems engineering references individual requirements as the starting points for a systems design.

Two standards are significant in the system engineer domain, and both can represent requirements models. The first, STEP AP 233 [9.2] represents systems engineering data including requirements but can be used to exchange data with nonsystems engineering tools, including higher level models such as those for the DoD architectural framework [9.3]. Second, SysML [9.4] is a UML profile designed for creating systems engineering models, and it has its own representation standard for exchanging between such tools. SysML modeling tools are designed to link the different systems models, and so provide an integrated environment. SysML exchange format will almost certainly be the basis for archiving SysML models, but AP 233 will be applicable to the output of specialist requirements modeling tools, and possibly for broader integration.

In March 2018, LOTAR announced it is to develop a part on model-based systems engineering [9.5 2018 Q1 & 2019 Q1 & Q3] and this likely to align with on STEP AP 243 "Modelling and Simulation information in a collaborative Systems Engineering Context" (MoSSEC).

9.3 **Parts Made from Composite Materials**

Composites—essentially mats of carbon fiber layered on each other and glued together—are used increasingly to build strong light structures, for example, wings. Once finished, a composite part cannot be taken to pieces, and is treated as a single part, with the shape modeled in a CAD system. The subject for archiving is the manufacturing design: the mats and the orientation of the fibers, how they are layered, and details such as inserts—noncomposite components that are built into the part, such as stress sensors. This manufacturing design, although using specialist CAD software, has as a major element the shape of the part, and the data model is common with that for mechanical CAD. A composite design will be extensively analyzed to see how it responds to physical stress and to other forms of damage, such as lightning strike.

Composite design and analysis was originally represented in STEP AP 209 [9.6]. Composite design has also been included in AP 242 [9.7]. The latter will have been influenced to meet the requirements of the LOTAR team developing archiving standards for composites. LOTAR is developing a number of parts on composites [9.8]:

- P300 fundamental and concepts for LT archiving and retrieval of advanced manufacturing
- P310 LT archiving and retrieval of CAD 3D composite design information

9.4 **Additive Manufacture**

Additive manufacture is better known as "3D printing," although it covers a number of techniques that produce a 3D shape without starting from a solid block and subtracting material from it. Additive manufacture is within the scope of AP242 [9.7], which will provide a data model for archiving. LOTAR kicked-off a working group in 2019 to support additive manufacture to create LOTAR parts in the range 350–390.

One concern for additive manufacture in aerospace is certifying the materials used and the parts manufactured. Although there are organizations that provide certification [9.9], certification can be expensive, so much so that some manufacturers isolate their machines from the Internet so they cannot receive the software creator's updates, as this may invalidate the certification.

9.5 **Finite Element Methods**

The performance of components under various environmental conditions—airflow, heat transfer, stress, vibration, and so on—is extremely difficult to solve exactly for all but the simplest of shapes. A finite element (FE) method uses a mesh to divide a surface or a volume up into small, simple shapes such as triangles or pyramids. It then formulates the analysis in terms of equations at each of the mesh's edges or faces, and solves the equations by a series of numerical approximations. The complication is that the answers for one mesh element depend in part on answers in the adjoining elements, and so FE systems solve all the equations across the whole mesh at the same time—think big parallel processing computers running for hours or even days at a time.

The results of an FE computation can be validated by physical experiment, for example, by placing a test shape in a wind tunnel. The advantage of FE simulations is that they can be used to predict the performance over many different configurations of an aircraft in many different environments. They become a target for archiving for two reasons. First, reuse: computations are slow and expensive; however, it may be possible to quickly estimate performance by interpolating from existing results. Second, certification and liability, where FE methods are used to justify that the aircraft or its support system meet certain safety criteria, for example, that the stresses on a component are within safety margins, or that the period between inspections is safe.

In some ways, the data is easy to archive, the model structure being a simple repetition of data for each element of the mesh. The most obvious challenge is the very large volume of the data, particularly because many FE methods run through a series of different software applications that pre-process, solve, and visualize the data, creating several datasets to be archived for each run. There are, however, two further issues. First, the software is often tied to a particular processor in order to make effective use of the way it breaks the problem into parallel streams. This makes it difficult to guarantee that the same starting data will give exactly the same results if run on two different machines.

Second, the validity of the codes, the mesh, the equations, and so on is demonstrated against experimental results, such as wind tunnel tests. Therefore, the test cases used and potentially the validation data may also need archiving.

FE data is represented in STEP, for example, in AP 209. LOTAR is working on this in its Engineering Analysis and Simulation working group [9.10], for which it has a memorandum of understanding with NAFEMS [9.11]—NAFEMS is the International Association for the Engineering Modelling, Analysis and Simulation Community. Three LOTAR parts are proposed:

- 600 Fundamentals and concepts for long-term archiving and retrieval of engineering analysis and simulation

- 610 Long-term archiving and retrieval of Simulation Process and Data Management (SPDM) information

- 620 Long-term archiving and retrieval of structural analysis information

This is also an area of considerable academic research. In addition, academic standards have been developed to cite experimental datasets.

9.6 Electrical Wiring and Circuits

If you look inside an aircraft before the internal paneling is in place, you will see bundles of wires running along the walls, often terminating in multiway connectors. These are the electrical harnesses or looms. The designers have to solve three problems: ensuring electrical circuits connect up the equipment, grouping the wires into bundles for installation, and physically routing the bundles through the available holes without bending the bundles too sharply. The last problem is typically solved using an add-on to a mechanical CAD system. Specialist electrical CAD applications are used for the first two problems.

The LOTAR team [9.12] are using STEP AP 212, AP 210, AP 239, and the VDA VEC specification [9.13] to develop archiving specifications for harnesses:

- 400 Fundamental and concepts for Long Term Archiving & Retrieval of electrical harness information

- 410 Long Term Archiving and Retrieval of physical electrical harness for design and construction

- 420 Long Term Archiving and Retrieval of CAD 3D electrical harness installation

There is now an Electrical Wiring Interconnection System (EWIS) domain under the CAx-IF [9.14].

9.7 Through-Life Support: PLCS

Much in-service information is continually updated as the aircraft is maintained, so it is not a target for archiving. However, the support solution includes the definition of maintenance tasks and the schedule for maintenance. These are developed in parallel with the aircraft design, but since the information is in active use through the life of an aircraft, it has not been identified as a priority for long-term archiving.

Product Life Cycle Support (STEP AP 239) (PLCS) [9.15] is an integrated model covering the "support solution" and the operation of that solution when the aircraft is in service. The AP 239 format has also been mooted as a suitable standard for metadata for the various archival packages [9.16]. This is due to the wide range of process data that lies within its scope and to its flexibility, which arises from its use of reference data. Reference data is a technique to allow data-driven customization of a data model, rather than fixing explicit entities and relationships.

9.8 Integrated Vehicle Health Management

Integrated Vehicle Health Management (IVHM) starts with sensors measuring stresses, temperatures, vibration, and so on, and building through a number of processing levels, identifying the current state of the aircraft and at the highest level, and predicting and scheduling preventive maintenance. Mostly, this data is downloaded at the end of each flight for immediate analysis. However, some types of data, such as the stresses that have occurred in flight, can be accumulated over the life of the aircraft and across the whole fleet. This data is used both to schedule inspections and replace parts, but it is also used to recalibrate the equations used in that scheduling [9.17]. This particular use case involves the storage of large quantities of data over a long period, and the maintenance of storage services over the same period, and was one of the drivers for the service architecture described in Chapter 5.

9.9 And since You Ask: Documents, Websites, and so on

Human readable file formats have been extensively researched by national libraries, academia, and so on, and there are many useful starting points ([9.18], see also [9.19, 9.20, 9.21, 9.22])—much of the research in this area is referenced as *data curation*. There are also tools available with functions ranging from identifying file formats to converting files to a common standard. Documents, websites, e-mail, and so on are not models in that they do not calculate new values if you change something, and in that sense, they are out of scope for this book. 2D geometry and drawings are explicitly outside LOTAR's scope, and it is assumed that for long-term preservation, they will be held as image files.

One significant difference between the archiving of documents and the LOTAR approach is that the various document and media standards do not come with validation properties other than basic data level tests such as checksums. However, the InSpect Project (Investigating Significant Properties of Electronic Content) [9.23] has worked on this issue.

9.10 Summary

This chapter has barely scratched the surface of the broader digital archiving landscape. In aerospace, much of the standardization efforts for data formats has focused on the STEP series of standards. Moreover, the LOTAR project has looked at many of the domains discussed and has published, or is working on parts covering them. Other, nonmodel formats, such as documents, have been extensively researched by other organizations.

References

9.1   Wikipedia, "Requirements Engineering," https://en.wikipedia.org/wiki/Requirements_engineering, accessed June 21, 2019.

9.2   ISO, "Industrial Automation Systems and Integration—Product Data Representation and Exchange—Part 233: Application Protocol: Systems Engineering," ISO 10303-233:2012, https://www.iso.org/standard/55257.html, accessed June 21, 2019.

9.3   Wikipedia, "Department of Defense Architecture Framework," https://en.wikipedia.org/wiki/Department_of_Defense_Architecture_Framework, accessed June 25, 2019.

9.4   SysML, "SysML Open Source Project—What Is SysML?," https://sysml.org/, accessed June 21, 2019.

9.5   "LOTAR progress reports," http://www.lotar-international.org/communication/progress-reports.html, lasted accessed November 25, 2019, referenced here by year and quarter, so [9.5 2018 Q1] is for the first quarter of 2018.

9.6   AP209.org, "ISO 10303-209: Multidisciplinary Analysis and Design—Welcome," http://www.ap209.org/, accessed June 24, 2019.

9.7   AP242.org, "STEP AP242 Project-Welcome," http://www.ap209.org/ http://www.ap242.org/, accessed June 24, 2019.

9.8   LOTAR, "LOTAR Composites Workgroup," http://www.lotar-international.org/lotar-workgroups/composites.html, accessed September 24, 2019.

9.9   Lloyds Register, "Providing Confidence in Additive Manufacturing," https://www.lr.org/en-gb/additive-manufacturing/certification-components/, accessed July 18, 2019.

9.10   LOTAR, "LOTAR EAS: Engineering Analysis & Simulation Workgroup," http://www.lotar-international.org/lotar-workgroups/engineering-analysis-simulation/scope-activities.html, accessed June 24, 2019.

9.11   NAFEMS, "Home," https://www.nafems.org/, accessed June 24, 2019.

9.12   LOTAR, "LOTAR Electrical Harness Workgroup," http://www.lotar-international.org/lotar-workgroups/electrical-harness.html, accessed June 24, 2019.

9.13   VDA (Verband der Automobilindustrie), "Vehicle Electric Container (VEC)—VDA-Recommendations," July 1, 2014, https://www.vda.de/en/services/Publications/vehicle-electric-container-vec.html, accessed June 24, 2019.

9.14   "The MBx Interoperability Forum," https://www.cax-if.org/ewis/ewis_introduction.php, accessed November 25, 2019.

9.15   ISO 10303-239, "Product Life Cycle Support," http://www.plcs-resources.org/ap239/, accessed June 25, 2019.

9.16   LOTAR, "LOTAR 'Metadata for Archival Package' Workgroup," http://www.lotar-international.org/lotar-workgroups/metadata-for-archival-package.html, accessed June 24, 2019.

9.17   Hebden, I.G. and Hunt, S.R., "Eurofighter 2000: An Integrated Approach to Structural Health and Usage Monitoring," presented at the RTO AVT Specialists Meeting on "Exploitation of Structural Loads/Health Data for Reduced Life Cycle Costs," May 11–12, 1998, Brussels, RTO MP-7.

9.18   Addis, M., "Digital Preservation: A Technologist's Perspective," presented at the Digital Preservation Coalition (DPC) Student Conference, London, 2018.

9.19   The Digital Preservation Coalition, "Digital Preservation Handbook," https://dpconline.org/handbook, accessed June 24, 2019.

9.20   Library of Congress, "Digital Preservation Formats," http://www.loc.gov/preservation/resources/rfs/index.html, accessed June 24, 2019.

9.21   Digital Curation Centre, "Home," http://www.dcc.ac.uk/, accessed June 24, 2019.

9.22   The British Library, "Digital Preservation," https://www.bl.uk/digital-preservation, accessed June 25, 2019.

9.23   "Investigating Significant Properties of Electronic Content," http://significantproperties.kdl.kcl.ac.uk/ — Final report at http://significantproperties.kdl.kcl.ac.uk/inspect-framework.html#_Toc252984221, accessed November 25, 2019.

CHAPTER 9

The Basics of PDM

10.1 What Is PDM?

The simple version of Product Data Management (PDM), as explained to a four year old: "An aircraft has a million bits, each bit has a drawing, and daddy's job is to make sure none of them gets lost." In the sophisticated version, PDM is the glue that sticks together customer requirements, assembly structures, parts and their models, and the aircraft production schedule. And it does this while everyone is changing their mind about what exactly is wanted and when. The sophisticated view takes a lot more explaining.

The focus of this chapter is the basics of *archiving* PDM. It works from a simplified design process that is representative of model-based design when it was superseding drawing-based design. A phrase also from that era is "islands of automation," and PDM was used to link separate design islands into an integrated process. Twenty years on, and the "integrated process" has become denigrated as the "stove-pipe system," and companies are looking to create linkages across the whole design organization, but that is a subject for the next chapter.

We start by describing how a small, integrated project team might use PDM to support their *design process*. *Process* is a key word, because PDM changes the perspective from archiving a *design model* to archiving *design process data*: for example, instead of the CAD model, the focus is on archiving the sign-offs that approve the CAD model. Three types of PDM data are archived: process data linked to design models, work orders (which control the design process), and product structure, which tracks the aircraft configuration.

To date, the LOTAR project has published a general introduction to archiving PDM data [L200]. This chapter and the next make use of these general principles, plus many of the discussions that contributed to them, but go beyond what has been published so far.

10.2 **The Role of PDM in Design**

In "Aircraft as a System-of-Systems" [10.1], a small design problem was teased out. The system designer sketched out a system to move a flap (Figure 10.1a). A mechanical designer took the sketch and produced a detailed design for a lever used in the system (Figure 10.1b). They worked with a stress engineer to develop a design that was both strong enough and light enough. A manufacturing planner then reviewed the design to ensure it was manufacturable. This section looks at this process from a PDM perspective.

10.2.1 **Configuration and Work**

Configuration control is the central process in aircraft design. Its objectives are to ensure that the configuration of an aircraft—the list of bits used to make it—is complete and accurate, and that the design of each bit has been properly checked.

FIGURE 10.1   Designing a lever.

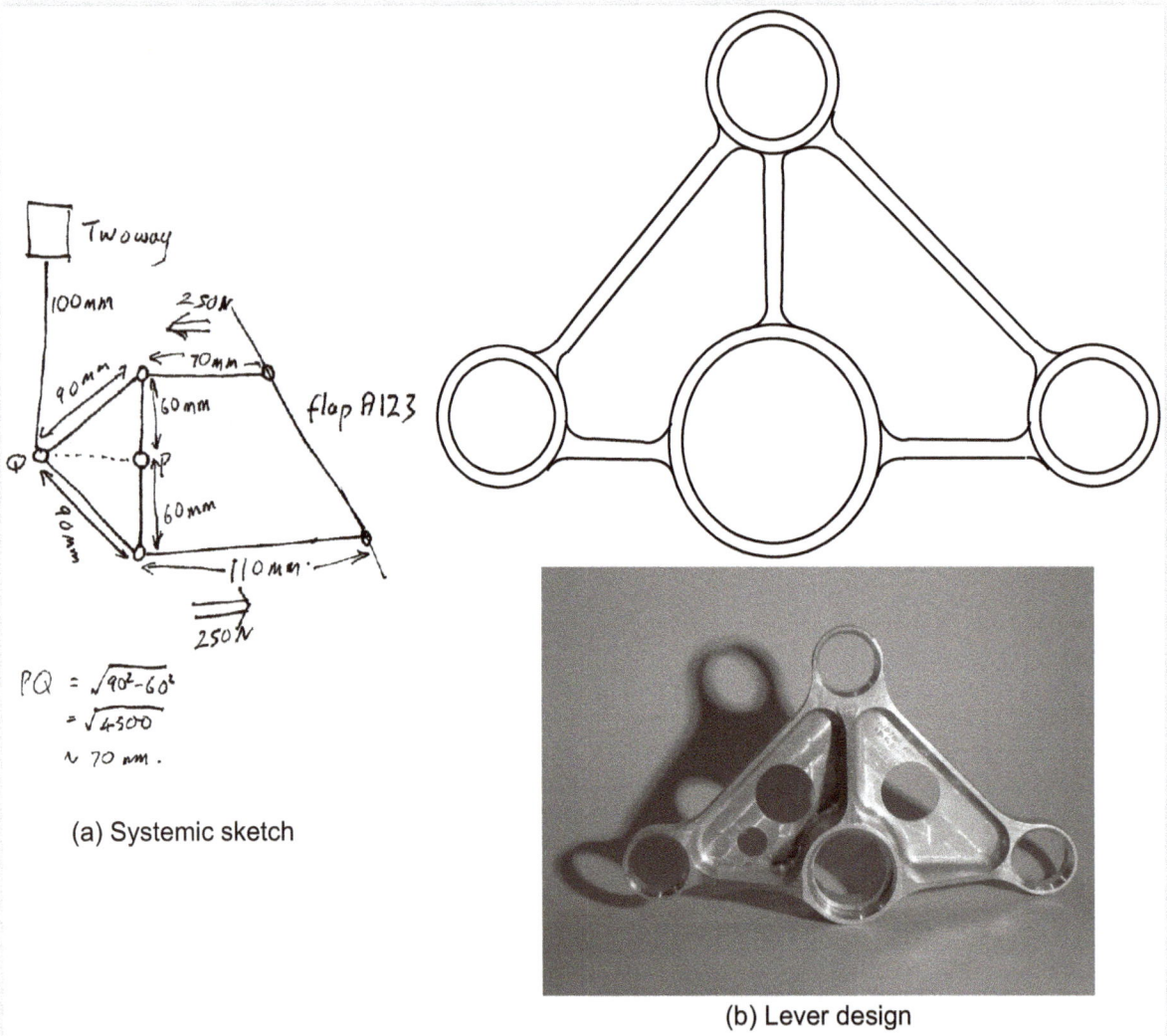

(a) Systemic sketch

(b) Lever design

FIGURE 10.2 Illustration of a *Work Order*.

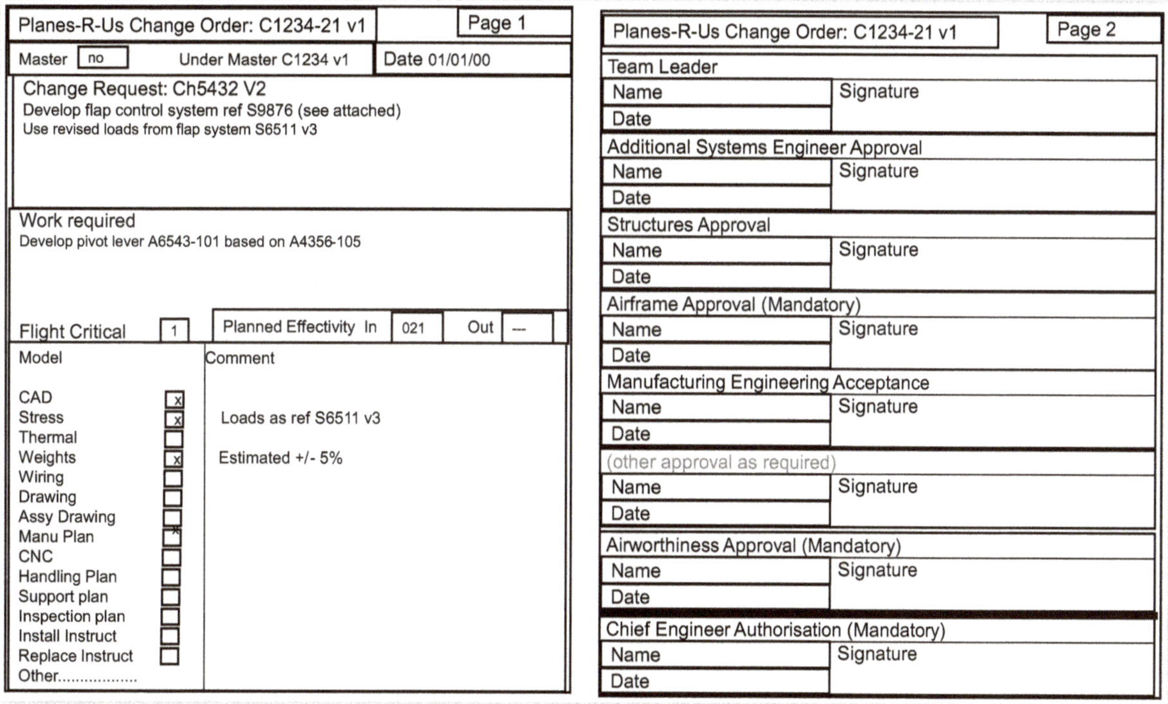

Planes-R-Us Change Order: C1234-21 v1		Page 1
Master no Under Master C1234 v1	Date 01/01/00	

Change Request: Ch5432 V2
Develop flap control system ref S9876 (see attached)
Use revised loads from flap system S6511 v3

Work required
Develop pivot lever A6543-101 based on A4356-105

Flight Critical [1]   Planned Effectivity In [021] Out [--]

Model		Comment
CAD	[x]	
Stress	[x]	Loads as ref S6511 v3
Thermal	[ ]	
Weights	[x]	Estimated +/- 5%
Wiring	[ ]	
Drawing	[ ]	
Assy Drawing	[ ]	
Manu Plan	[*]	
CNC	[ ]	
Handling Plan	[ ]	
Support plan	[ ]	
Inspection plan	[ ]	
Install Instruct	[ ]	
Replace Instruct	[ ]	
Other..................		

Planes-R-Us Change Order: C1234-21 v1		Page 2
Team Leader		
Name	Signature	
Date		
Additional Systems Engineer Approval		
Name	Signature	
Date		
Structures Approval		
Name	Signature	
Date		
Airframe Approval (Mandatory)		
Name	Signature	
Date		
Manufacturing Engineering Acceptance		
Name	Signature	
Date		
(other approval as required)		
Name	Signature	
Date		
Airworthiness Approval (Mandatory)		
Name	Signature	
Date		
Chief Engineer Authorisation (Mandatory)		
Name	Signature	
Date		

The lever is a *Configuration Item* (CI), an identifiable element in the design-in this case a physical part. Under standard configuration management rules [10.2], a CI can only be created or modified under the authority of a *Work Order*—a simplified example of a paper work order is shown in Figure 10.2.

Details to note include:

• The *Work Order* is a consequence of a Work Request, here for a design update.

• The new part is given its own configuration identifier (its part number)—although the design is based on an existing part, the two parts differ in fit or form or function (or all three), and so, under configuration rules, must be given a new part number.

• There is a long list of design models that may be required for a part, although in this example, only four will be created—the CAD model, the stress assessment, the manufacturing plan, and the estimated weight. The CAD model will be stored in its own file controlled by the PDM system, as will the stress analysis and the manufacturing plan. Estimated weight is calculated from the CAD model, but the result will be stored directly in the PDM system because, to find the overall weight for the aircraft, a separate application will add up the weights of each component and it is impractical to do this by opening every CAD model.

• A long list of sign-offs is required before the part is issued. Some are required to show the part is airworthy, others ensure the part is cheap to manufacture and support.

• "Planned Effectivity" identifies where in the production schedule the new part will replace the old design—see Section 10.2.3 for details.

• Finally, a senior engineer signs the Work Order off as being complete and the new design is formally issued.

CHAPTER 10

The *Work Order* provides the record of the upper layer of the design process. It links together what the customer wants (the *Work Request*), the project plan (which sets the start date for the work), and the production schedule (setting the date when the new design should complete and can be used to build an aircraft).

10.2.2 The Configuration Item

A CI is a design for an aircraft component. What counts as a CI and the rules for identifying them are part of the *Configuration Management Plan* [10.2].

A PDM system automates the design stage of a configuration management system. It does three things with a CI: it holds the master record for the CI; it manages the models that define a CI, preventing unauthorized change; and it animates the change process as a workflow, recording any process data (e.g., an approval) and sending it to the next person in the process.

A *Work Order* will be created in the PDM system by the Configuration Management Team, and they will use a workflow to send it to the design team. On receiving the *Work Order*, one of the design team will fill in a CI template to create a new Configuration Item. The first template tab—the identification tab—will show the CI is of type "part" and give it a part number. Other CI types include assembly, system, and software module. The tab will also hold a description, any alternative part numbers (e.g., NATO Stock Number), plus a link to the *Work Order*. These fields, as well as being archived with the CI, will be indexed as search terms within the archive.

Other tabs will be filled in as the design progresses. In this example, a tab will be created for each independent design model: the CAD model, the stress model, and the manufacturing plan. Each tab will explicitly record the model identifier, although the PDM system will directly control the physical files for the model. The tab will also record the status of the model (*in-work*, *in-review*, *approved*, *cancelled*) and record its sign-offs. Once a model is *in-review*, the PDM system will prevent further change unless it is kicked back to *in-work*. These model tabs are part of the PDM system's *design management* function, but archiving is concerned only with the end result, when the models are *approved*.

There will also be an effectivity tab (see Section 10.2.3) and an "odds-and-ends" tab—a place to hold values that either do not fit into any particular model, or which would be impractical to access otherwise. An example of the former is a "large part flag," set to warn manufacturing that they need a crane when moving the part. The estimated weight is an example of the latter.

The final tab is the "sign-off" tab. The sign-off process for CI involves a review by a number of specialists, as illustrated in the list of signatures required by the *Work Order* (Figure 10.2). These will not only check that design rules have been interpreted correctly, but also that the intent of the design rules has been met. Once all the specialist reviews have taken place, a sign-off by Air Worthiness is needed (the part **may** be fitted to the aircraft) and finally the engineering authority *issues* it, showing the change is complete and **should** now be embodied in the aircraft.

The CI design package is then "released." In this example only four models are included, however, a release may include 30 or more different elements, such as manufacturing drawings, analyses, and support plans and procedures. Historically, this was the point at which the design was archived.

One of the characteristics of a PDM system is that it is easily customized to match up to the customer's business process, so that each tab may have its own custom fields, and the field names and values customized to match the process—for example, "revision" or "issue" with values running 1, 2, 3, … or A, B, C, …. This means that any exchange of PDM information starts by mapping the current customization to whatever interface standard is used, even if the target is another copy of the same software-an illustration of the fact that *information* is a mapping of the *business context* to *data*.

10.2.3 The Product Structure

The best-known example of a *Product Structure* is the Bill of Materials (BOM), a tree whose root is the whole aircraft, whose intermediate nodes are stage builds and assemblies, and whose leaves are the parts. As the simple lever is a single part, it does not have a BOM. A simple BOM (also from reference 10.1) is illustrated in Figure 10.3. It is shown both as the tree for an assembly and as its text equivalent, an indented list. In the tree, the line between nodes is the relation "is part of." These mathematical "trees" are always shown with the root at the top—in the figure it is "Section AJ27-701."

The parts, the assemblies, and the aircraft itself are individual CIs, and these are the anchors for both design models and design process data. For example, the CI for an assembly holds the assembly model (these parts in these positions) and the approval that shows "these parts put together in this way" is airworthy. The Product Structure therefore defines the configuration of the aircraft.

That statement must be qualified immediately. First, there are multiple product structures, each with a different purpose. For example, a hydraulic system may run between multiple assemblies, but one of its key properties is that it does not leak hydraulic fluid. This cannot be verified by testing one assembly at a time but only by testing the hydraulic system as a whole [10.1, chapter 5]. Consequently, there is a *systems* product structure to record systems-level data and approvals. The structure discussed here is

FIGURE 10.3 Bill of materials.

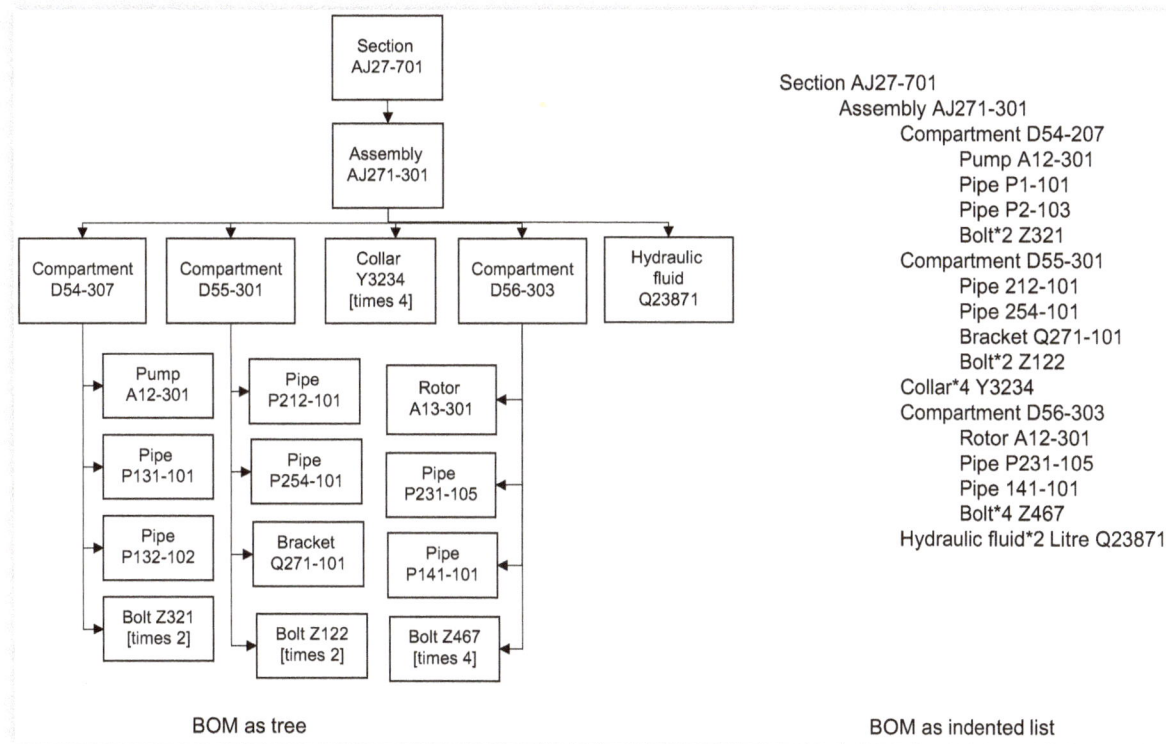

Section AJ27-701
   Assembly AJ271-301
      Compartment D54-207
         Pump A12-301
         Pipe P1-101
         Pipe P2-103
         Bolt*2 Z321
      Compartment D55-301
         Pipe 212-101
         Pipe 254-101
         Bracket Q271-101
         Bolt*2 Z122
   Collar*4 Y3234
   Compartment D56-303
      Rotor A12-301
      Pipe P231-105
      Pipe 141-101
      Bolt*4 Z467
   Hydraulic fluid*2 Litre Q23871

BOM as tree   BOM as indented list

the *design* product structure, and since it is the structure used for certification, it is the initial focus for archiving.

Second, each *Work Order* changes the configuration, updating the design of individual parts, and replacing some parts with new designs. In small-scale design projects, this would lead to a new version of the product. Aircraft are not small-scale and have many changes running in parallel. *Effectivity* is the way these parallel changes are managed. An *effectivity* is a statement that "this CI (part, assembly, etc.) is used in aircraft numbered M onwards"—M is the tail number based on where the aircraft appears in the product schedule. When the team design their part, the *Work Order* identifies the planned effectivity of the part they design.

However, an assembly explicitly lists every CI that it is made from. To use the new part, the assembly tree has to be updated so that the old assembly "cuts out" at aircraft M-1 and the new one "cuts in" at aircraft M. This update creates a new version of the assembly. This change in turn needs propagating to update the next higher assembly. The exact way effectivity propagation works is one of the mysteries of the aircraft world.

One important function of a PDM system is to filter a product structure by *effectivity*, thus giving the configuration of one particular series of aircraft—or perhaps just one aircraft. This is also referred to as a "resolved product structure." Conversely, the "where used" enquiry can be used to identify every aircraft where a particular part is used, and the query is needed, for example, in the assessment of change proposals.

10.3 Business Processes and Data

10.3.1 Process and Data Structure

A PDM system manages the work and configuration processes, so the data recorded depends on the process. A striking example of this comes from the 1990s when model-based processes replaced drawing-based processes.

In the drawing-based process, a drawing could show multiple parts, so when a Work *Order* gave permission to update a **drawing**, it implicitly gave approval to change multiple parts. In the model-based process, the *Work Order* gives approval to change the **model** of a particular part, and subsequently to produce a new drawing of that part. Note that in normal conversation, the CAD model would be referred to as a "part." The information structures involved are shown in Figure 10.4. The upper diagrams show the abstract data model, and the lower show what happens when the data model is populated with individual parts and drawings.

The figure is based on the EXPRESS data modeling convention [10.3], in which each box represents a type of entity: *part*, *drawing*, and *Work Order*. A relationship is shown by a "lollipop," and it goes from the entity holding the handle to the entity on the sticky end. A subtlety of this convention is that the entity at the sticky end can exist independently without there being an entity to hold the lollipop. That is, in the drawing process, there can be a *drawing* before any *part* is defined, but no *part* without a *drawing*, while in the model-based process there can be a *model* without a *drawing*. The two data structures are incompatible, and data from one process would be rejected as erroneous by a system based on the other. This would make it impossible to load data from a drawing-based process into a modern PDM system.

The move from a drawing-based to a model-based process involved the sort of fundamental process change that happens rarely—partly because it entails replacing all the existing PDM systems (millions of dollars). More mundane are revised sign-offs

FIGURE 10.4   Information structures for (a) drawing-based and (b) model-based processes.

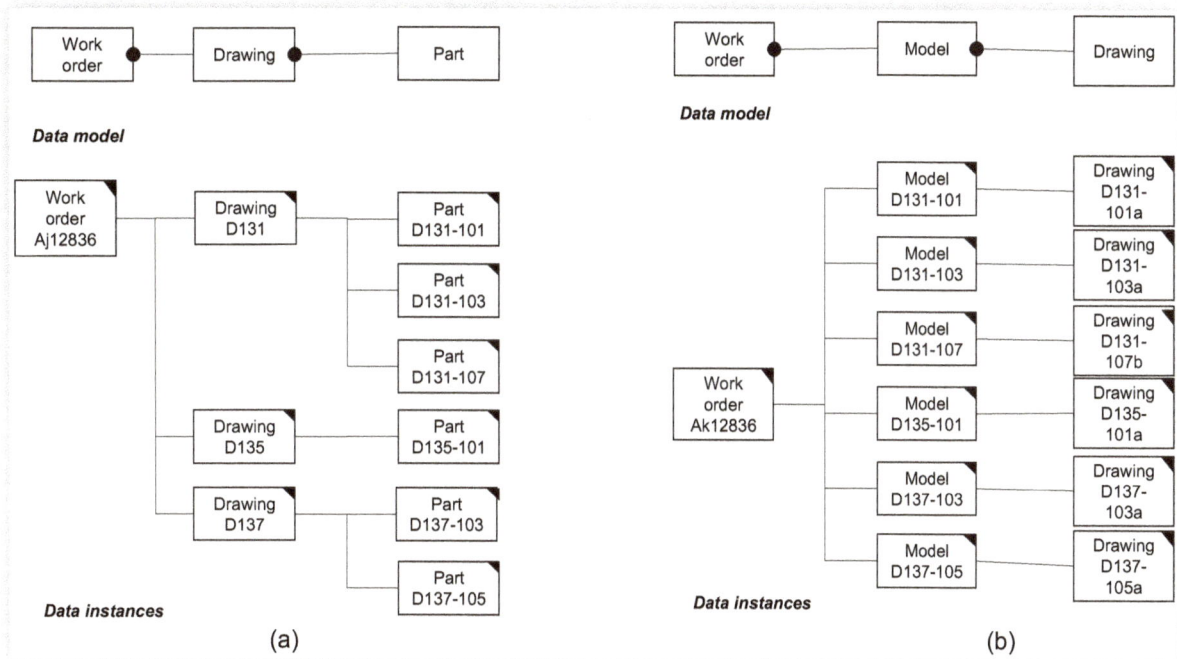

list, for example, adding a lightning-strike sign-off. LOTAR requires that the impact of any change to the PDM system is assessed against the need to access archived data.

10.3.2 Process and Meaning

A conventional approach to meaning is that a word represents an idea. Without going into a philosophical discussion, this is not useful when dealing with industrial systems, particularly when part of the system is a computer—computers are not known for their grasp of ideas. For a less challenging approach, allow the meaning of a term to be given by the behavior of the system which interprets it. For example, a computer interprets the term "secret" by testing whether a user has permission to read a document marked "secret," and it has no other concept of what "secret" means.

In business, the process sets the behavioral rules that apply to the actors in the process. For example, the possessor of a "secret" document must record it in a register, must not photocopy it, and must lock it away when not in use. The exact details of "secret" are defined by the organization setting the rules—and in the 1990s, UK and US rules differed. UK "secret" corresponded to the US "top secret," while the US "secret" corresponded to UK's "confidential"—significant confusion ensued. Terms like "secret" allow humans to use the "word = idea" as a shortcut, but the precise meaning is given by the behavior of the system.

PDM process terms such as "approved," "authorised," and "issued" are best treated in this way, as their exact significance is defined by the process. For example, a design might be "approved" for release to suppliers—for example, to show where their parts fit in a test rig—or "approved" as airworthy and able to be used in a production aircraft. If the designated community is to understand the data when retrieved, they must understand the process under which the data was created. This requires that the process documentation is archived and referenced by the CI's *preservation description information*.

10.4 **Standards and PDM**

CM processes in aerospace broadly follow EIA 649 "National Consensus Standard for Configuration Management" [10.2], although this provides only the overall structure, not the detail such as the departmental sign-offs. The data generated in these processes is the basis for the STEP standards. For PDM, the first industrial-scale, electronic data exchange [10.4] used the STEP PDM schema [10.5]. Its scope is illustrated by its "units of functionality" including: Part Identification, Part Structure and Relationships, Relationships Between Documents and Constituent Files, Authorization, Configuration and Effectivity Information, and Work Management. The PDM schema is built into the main STEP APs. The PDM-IF (Implementers Forum)[1] works to ensure that standards supporting PDM—particularly AP 239 and AP 242—are fit for purpose and consistently interpreted. They also test implementations of these standards.

10.5 **Archiving PDM**

10.5.1 **Basic Principles**

For one reason or another, the aircraft configuration continually evolves, with the design being updated well into the in-service phase. Consequently, there is no "finished" PDM, and PDM data needs to be archived throughout the aircraft design cycle. This requires that PDM data is broken into discrete submissions, but what are the Submission Information Packages? and how do we maintain the links between the contents of each SIP and the configuration as a whole?

There are four distinct lifecycles within PDM:

- Work management, where each piece of work follows a direct creation-to-completion path.

- An individual design model, which follows a series of *in-work* to *approved-for-use* cycles.

- A CI, which may be updated by a series of work management cycles, with each update being a container process for the cycles of the individual design models.

- A product structure tree, which continually evolves as its CIs are updated.

In the model-based process, as described above, the lifecycle of the CI and the lifecycle of the individual models coincide, so that the sign-off of the whole CI also involves the sign-off of the contributing design models.

One principle for archiving is that where a submission refers to something external, that external item should be in the archive already, and its presence is checked by the submission process. In the design process described, CIs refer to the various component models, while the models refer back to the CI as their defining context. Moreover, the CI provides the context that links the stress model to the CAD model. Archiving the CI and the models together avoids the mutual referencing needed if the CI and the models were archived separately. However, the context for the CI is a *Work Order*, which in turn

---

[1] Reference data is a technical subject in data modeling. A reference data system is customized by changing the data rather than the software.

points back to a *Work Request*. Thus, the problem of referring to items not yet archived remains, and it would be better to solve this conundrum and archive the CI and the models separately. This is discussed further in the next chapter.

A second principle is that the archive should be simple to operate and inexpensive to use. For PDM, this means keeping the complexities of PDM in the PDM system, and not duplicating them in the archive. For example, "Where Used," searches the Product Structure for instances of a CI, but an archive would not search inside an AIP that contains product structure, nor would the archive do *effectivity* filtering.

10.5.2 Archiving Models

The details of archiving the model are given by the appropriate LOTAR part, such as part 110 for 3D CAD models. The models will be archived with limited process data: in the LOTAR part for CAD models (Section 8.2.4), this is limited to data about the submission process. The approvals are held by the CI, which is referenced by the Part Identifier and Version. The model identifier and issue can be used to resolve references from the CI to the model.

10.5.3 Archiving Work Documentation

A *Work Request* and its *Work Order* record one step of the aircraft development history and provide the link between the modification the customer asked for and what is actually delivered. LOTAR plans to cover this in part 250 "Change documentation" (historically *Work Requests* were called *change requests*).

The essential data for *Work Request* is what is requested, who requested it, and the list of *Work Orders* that are scheduled from it. The *Work Request* will be signed off once all the scheduled *Work Orders* have been signed off.

The essential data for *Work Order* includes the request that scheduled it, the CIs referenced, the CI effectivities, the work descriptions, and at least the final approval, if not the whole set of signatures applicable. The data will be archived once all the CIs have been signed off and the *Work Order* receives its final approval.

Work items may be viewed as structured documents which are held in PDM as complex data structures. They can be directly represented in protocols such as the STEP PDM schema, but the PDM schema does not cover business-specific customizations such as an extended list of approvals. Since completed work is historical data, the case for data reuse is not obvious. A simple document format—say SGML or HTML—may provide an easier route than using the PDM schema, which was designed for work-sharing between businesses. The format would be read-only, which should satisfy use cases for certification or legal access. Legal use would require proof that the process creating the document both reliably extracts the data and reliably maps it to the document format. Using a document format would not require recustomization of the PDM system to read archived work items.

The signatures in work activity data raise two issues. First, the signatures will be generated electronically, but does the archived data simply record the fact of the signature (who and when), or is a copy of an encrypted digital signature required to prove authenticity? Second, in the long term, the signatures will require supporting context, showing at least that the signer was a specific employee and that they were suitably qualified to sign. This creates two subproblems: how to archive the personnel details? and how personal data protection legislation applies?

10.5.4 **Archiving Configuration Items**

The essential data for the CI is its identification data, its links to the *Work Order* and to the design models, the approvals, and any design significant data in the odds-and-ends tab. The core data model is covered by the STEP PDM schema; however, this will need extending to cover the full list of sign-offs. One possible alternative is the PLCS schema, since that covers the same scope, but is more reference-data driven [10.6]. Validation rules are likely to include the number of models and hash codes generated from model identities. The CI would be archived once the new version of the CI is formally issued.

10.5.5 **Archiving Product Structure**

At any one time, multiple changes are in progress, each of which affects a different part of the aircraft and will be completed at a different time. Effectivity allows these changes to proceed independently, but if we ask "what is the actual product structure" we will get different answers depending on where in the production schedule we look, and whether we look only at closed changes or include changes in progress. Or, rather like a quantum world, the PDM system holds many different product structures in superposition and gives a single tree only when we make an observation on it.

This leads to the question, "What is archivable?". There is no requirement to archive work-in-progress [L200 2.2], and so we can exclude any potential future product structure that is the subject of an open *Work Order*. The linkages in the structure between CIs are conditional on effectivity. That points to three possible sets of data, called here a snapshot, a baseline, and a resolved product structure. They differ in the effectivity rules used to extract them. In addition, to reduce the amount of data archived at one time, these sets can be divided into subtrees to be archived separately, with the leaf of a higher-level subtree pointing to the root of a separately archived lower level subtree.

A snapshot copies the contents of the PDM system at the date it is extracted, and no effectivity filtering is applied. It will include the different variants and completed changes even if they are predicted to cut in later in the production sequence. It may include designs that have already been superseded. To examine any particular aircraft configuration, the snapshot will need to be loaded into a PDM system to resolve effectivities.

A baseline is a description of the configuration at a point in time, typically at a significant point in the project, such as the transition from prototype development to production design. This can be translated into a tail number (M, say), and the effectivity filters set to extract the product structure for M, in which case the CIs selected are the ones effective for that point, but excluding any changes planned to cut in later or have already been cut out. The baseline will include all variants active at the time.

A resolved product structure is the product structure applicable to a particular aircraft, with other variants removed, but incorporating any changes that are signed off and effective. Resolved product structures for different variants will overlap, for example, where the designs have stage builds and assemblies in common. It will need archiving in subtrees to avoid duplicating data in the archive. One use of a resolved product structure is to extract as-manufactured and as-maintained configurations, so that additional information—for example, wavers, limitations, life-expired—can be attached to follow the individual aircraft as it progresses through its life.

The essential information for a product structure is the root CI and the list of children for each CI together with their effectivities. The core model is common to many STEP schemas, including AP 203, AP 214, and AP 242. LOTAR has not yet published any verification rules, but it has demonstrated validation properties that can detect when parts of the product structure are missing or misplaced. Industrial data exchanges [10.4] also demonstrated that it is possible to maintain a product structure within a PDM system by a series of incremental updates (changes since the last exchange), with the PDM system stitching the product structure back together.

Reading an archived product structure will probably require the use of a PDM system to resolve effectivities, except for the case of a resolved product structure, for which a simple product structure viewer could suffice.

The LOTAR parts for archiving product structures are yet to be published. Separate parts are planned for different stages in the lifecycle:

- 210—As Designed
- 220—As Planned
- 230—As Delivered/maintained
- 240—In-development (i.e., records from the early development phase, e.g., prototypes)

10.6 **Summary**

Configuration management and product structures aim to ensure that the configuration of any given aircraft is certain, while allowing many teams to work in parallel. Exactly who is involved and what they do depends on the company business process, and this in turn defines the content and the meaning of an archival submission.

PDM archiving covers configuration management and product structure, and is more focused on maintaining links between design artefacts than archiving individual artefacts. Historically, the design view was identified with the CAD models of the physical parts and assemblies. However, a modern Integrated Design Environment will pose new challenges, which are discussed in the next chapter.

Reference

10.1 Barker, S., *Aircraft as a System-of-Systems: A Business Process Approach* (Warrendale, PA: SAE International, 2019, ISBN-Print: 978-0-7680-9402. Chapter 6 covers the technical side of Product Structures in more detail (though long-term archiving gets only two pages).

10.2 ANSI/EIA-649B, "National Consensus Standard for Configuration Management," TechAmerica, accessed May 22, 2019, https://archive.is/20120801125835/http://www.geia.org/ANSI-EIA-649-A-Standard----NATIONAL-CONSENSUS-STANDARD-FOR-CONFIGURATION-MANAGEMENT.

10.3 Wikipedia, "EXPRESS (Data Modeling Language)," accessed June 19, 2019, https://en.wikipedia.org/wiki/EXPRESS_(data_modeling_language). See also ISO 10303-11 1994 or 2004.

10.4 By the Eurofighter consortium (UK, Germany, Italy, Spain), supported by PDES Inc. and ProSTEP c. 1999-2001.

10.5 PDM Implementor Forum, "Usage Guide for the STEP PDM Schema V1.2," Version 4.1, January 2002, https://www.prostep.org/fileadmin/downloads/ProSTEP-iViP_Implementation-Guideline_PDM-Schema_4.3.pdf, accessed May 30, 2019.

How to Archive an Aircraft

11.1 Problems of Scale

If you want to talk to the engineers who design an aircraft, the three key enablers are a map, a car, and a list of phone numbers. An aircraft factory has many buildings, and you need a map to find the concerned department. The buildings are large and spread out, and you may need to drive 20 minutes or more to get to the next department. And when you get there, you will need to ring for a guide to get you through the labyrinth of cubicles and desks. The problems of aircraft design are often problems of scale.

Most of this book has focused on one particular problem—the short timescale for software updates compared to the long timescale of an aircraft lifecycle. This chapter looks at a different scale problem, the difference between how much design a single team can do compared to how much goes into an aircraft. Aircraft design involves thousands of engineers, and even a straightforward design, such as a new variant of an existing aircraft, can take 2–3 years to complete.

This gives rise to two issues for archiving. The first is obvious—how do we archive a design a bit at a time.[1] The previous chapters looked at archiving specific bits—CAD and other models, CIs, work activity, and product structure. But the last chapter skirted round the issue of links between the different bits, and this is dealt with in Section 11.2. The second issue arises from the aircraft companies' desire to reduce costs and time to market. Their response has been to work toward an integrated design environment (IDE). The technology is summarized in Section 11.3, but more significant are the process changes that it will entail, since these will have an impact on PDM archiving.

---

[1] In the context of PDM, the term "incremental archiving" generally has the sense of "archiving the changes since the last increment," rather than "bit by bit."

11.2 **Gluing Things Back Together**

A key function of PDM is to glue everything together. It sticks the work activity to the CIs, and the CIs to the models. It also sticks the CIs together into product structures. What PDM implements is a continually evolving database of links between the items. One option might be to keep the PDM system running for the life of the aircraft, but to date these systems have had a relatively short lifecycle, depending on the business plans of the software suppliers, not the lifecycle of an aircraft.

The alternative, as discussed by LOTAR, is to archive fragments. This requires preserving the links between fragments. This raises three questions:

- What are the fragments?

- How do we preserve the links?

- How do we glue things back together when we take them out of the archive?

The sketch below points to some of the main possibilities.

11.2.1 **What Are the Fragments?**

Give an engineer a design problem, and their first step is to learn the problem, after which they try to learn the solution. And that learning takes time and effort and is factored into the cost of design. The best time to archive is shortly after the design has been completed, just as the best time to write the minutes is shortly after the meeting. Wait too long to write the minutes, and some of the details fade and get confused; fail an archival quality check and the design will need revisiting, so if you wait too long to do the check, the cost goes up as the engineer relearns the design. And once the design is signed off, the engineer will move on to the next design, and learning the next design will further crowd out the last. Moreover, efficiency also requires archiving checks to be part of the approval process (see Figure 5.4), so from this perspective there is no advantage to deferring archiving. Therefore, the choice of fragments to archive is driven by a combination of human factors and the lifecycle of the design items.

The previous chapter identified separate lifecycles for *Work Requests*, *Work Orders*, CIs, design models, and product structures. Moreover, each model and item type has its own submission process, for which specific key characteristics, core model, validation and validation properties, and metadata should be specified.

A second consideration is that these design items are directly significant in themselves and will be the targets for retrieval in a future process. Consequently, the business already has a way of referencing each of them, either explicitly—"Part number A123-101 version 1"—or by context "The CAD for model Part number A123-101 version 1." Yes, the business also can reference at a finer level of detail, for example, "Surface S103 in the CAD model for Part number A123-101 version 1." However, a surface will not be accessed directly from the archive, but by pulling out the CAD model and using CAD software to find it within the model.

The best assumption is therefore that each item detailed in Chapter 10 has its own submission process, its own SIP, and its own AIP. It might be desirable to archive a group of items at the same time, for example, the release package for a CI, which contains all the models and the CI process data. However, the submission process would then have to split the models out for their own model-specific checks, and a quality failure of any one of the models would mean that the submission as a whole would fail to complete, and the whole would need resubmitting.

11.2.2 How Do Items Reference Each Other?

The direct links between items in PDM occur at two levels. The first follows from the old manual processes, using an explicit item number—the *Work Order* number, the *part number*, the model identifier, and so on. When the PDM items are submitted, the item numbers by which they are referenced will be indexed against their AIP number within the archive's own data management system, so the item number can and will be used as a search term for the archive. The second link is internal to the PDM system and may be implemented as the PDM system's internal object reference, some form of hyperlink, or a contextual link, such as from the model tab for a CI. The PDM system uses these links to speed up navigation through the PDM system interface, and they have no meaning outside of that specific PDM system, so will not be archived.

Over the long term, new types of models will be developed, the PDM system will be replaced, integration environments merged, business processes will be revised, and businesses will be restructured. These may compromise the uniqueness of business identifiers. Only the archive will endure (though that may go through several software implementations). One risk reduction option would be to export the AIP identifier back to the PDM system to be stored with the item identifier. Then when another item is submitted, its references to the first item can include the AIP number, and this could take precedence over the item identifier if it is found to be corrupt.

11.2.3 Forward and Backward

The issue for archiving fragments is that they are added sequentially. A CI references the models it contains, and the model references the CI to provide its context. If a model is added before the CI, then the context for the model does not exist in the archive, while if the CI is archived first, the existence of its models cannot be checked. This conflicts with the requirement that referenced items should already be in the archive.

There are three possible ways out of this bind. The first is to submit both items at once. However, this breaks away from the timing criterion (once a design item is complete) that informed the identification of the archiving fragments. Moreover, the chain of reference continues backward from the model to the CI to the *Work Order* to the *Work Request*. A *Work Request* may cover a major change, such as a new wing design, and we are back to the problem of trying to archive everything at once.

The second is to use the business identifier as the cross-reference, omit the existence test on submission, and expect that the referenced item will be submitted eventually. However, that submission may not happen for a number of reasons, from error to the change being cancelled. Without the ability to follow the reference, the engineer of the future may be left scratching their head, wondering what the design was all about.

A third option is to divide the references into forward and backward references. A forward reference is one to something already in the archive, and a backward one to something that will be submitted in future. On the analogy of STEP's *Promissory_usage*, for the items that are backward referenced, a "promissory" dummy submission is made when the item is created, and these become the target for the backward references. Once the actual item is submitted, then the reference to the promissory item is updated to that of the actual item. The submission sequence is shown in Figure 11.1 (greyed out items are parallel sequences where a *Work Request* can be fulfilled by several *Work Orders*, etc.).

A submission can then check that all its references can be found in the archive, although some will be "dummy" promissory items. The archive itself will then run

FIGURE 11.1   The submission sequence for promissory items.

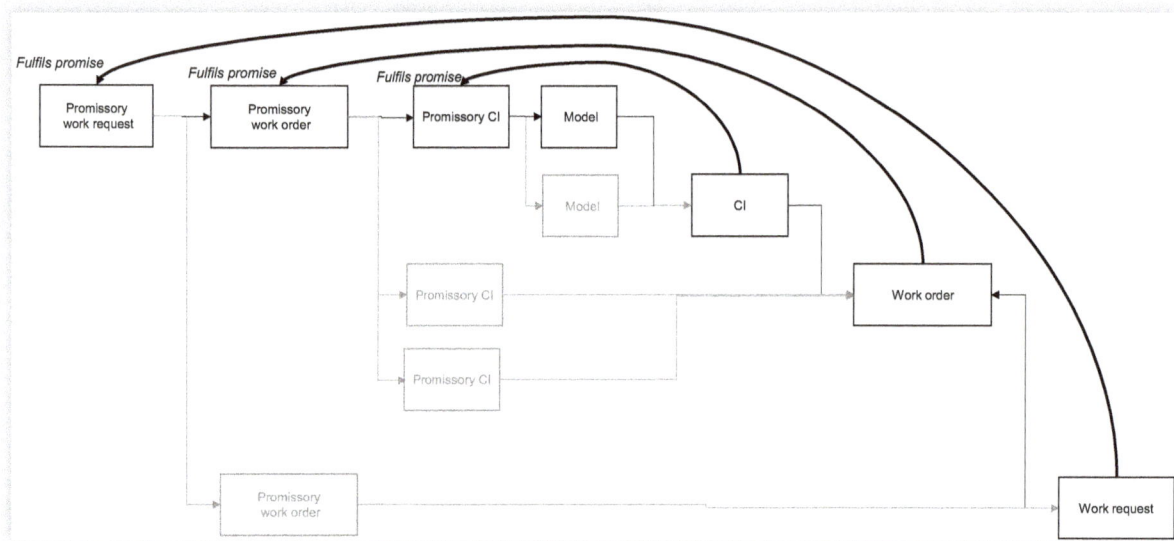

periodic checks to find promissory items and to follow up on ones whose submission appears to be significantly delayed.

11.3 Integrated Design Environments

11.3.1 What Is an IDE?

Computer hardware has gone through a number of evolutions from stand-alone machines, through clusters to networks. Technologies such as object brokers [11.1] were developed to allow software to follow a similar evolution, but much software is still built on the stand-alone model of input-process-output. Nevertheless, a great deal of integration has been achieved through the use of wrapper software. PDM systems illustrate one such approach, in which the PDM system focuses on controlling input and output files and providing workflow, but leaves the processing to specialist applications. Engineers' workbenches provide another approach, which focuses on taking the output of one program and piping it to the input of another, so automating a complex processing sequence using several specialist programs. An IDE provides step up in the level of integration.

An IDE can federate a number of separate departmental design environments, each of which may use their own specialist workbenches. It provides data management at the file level and links the departments through workflow, rather like a PDM system, but puts much more emphasis on coordinating work-in-progress. At the enterprise level, it coordinates design cycles, informing a department when inputs from other departments are ready. At the department level, it coordinates teams. The aim is an improvement in situational awareness [11.2]—a combination of knowing what everyone is doing, how far they have progressed, and what are their plans for the next step. It also speeds up the design process by both automatically notifying people when a process step has finished, and making the resulting data directly available, rather than waiting on someone to send it on.

11.3.2 The IDE Process and Its Information Model

Design review criteria generally include cost of manufacture and support, so this data should be linked to the CI. Since an IDE is used to coordinate the development across all departments, an IDE can link these views to a CI—rather like a PDM system with many more model tabs for each CI.

An update to a design is a configuration change, which must be formally reviewed. Moreover, the change not only creates a new version of the CI, but this change must be propagated up the product structure. Configuration changes are a significant part of the cost of a design.

Changes to manufacture and support can occur throughout the service life of the aircraft. Provided there is no change to the fit, form, function, or service life, a new manufacturing process should not require a new version of the design and does not require the reapprovals that a design change implies. Similarly, a change in a support solution-to use a new inspection method or a revised replacement procedure-does not change the design.

The standard PDM data structure (Figure 11.2a) can be used to group design, manufacture, and support models into different views. However, these depend on a single-version entity for the CI, and so a change in any view creates a new version of the CI. A modified data model has been proposed [11.3], which adds a view iteration counter to each view (Figure 11.2b). This counter can provide the release control for each view and allow each view to be released separately. Only when there is a change that affects more than one view is it then necessary to create a new version of the whole CI. Figure 11.3 shows the configuration lifecycle in terms of this data model.

The view construct has been part of the STEP model since its inception. However, in the initial adoption of a model-based process, the design was identified with the CAD

FIGURE 11.2 Data structures for PDM and IDE.

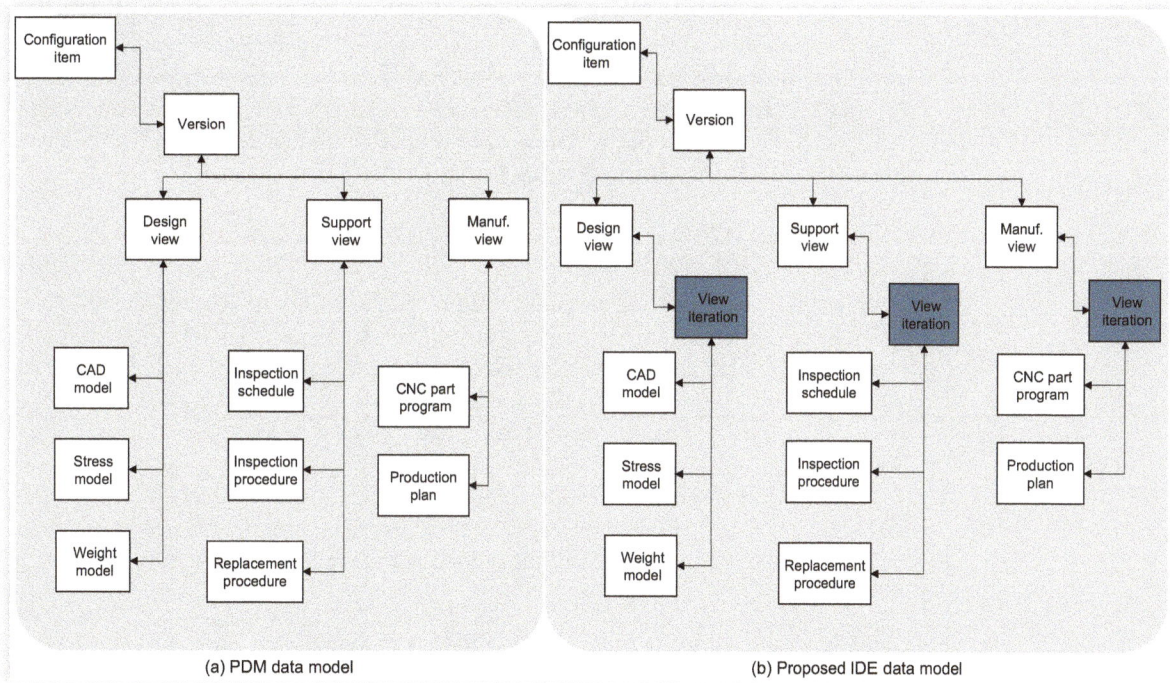

(a) PDM data model   (b) Proposed IDE data model

CHAPTER 11

FIGURE 11.3 Changes using an IDE data structure.

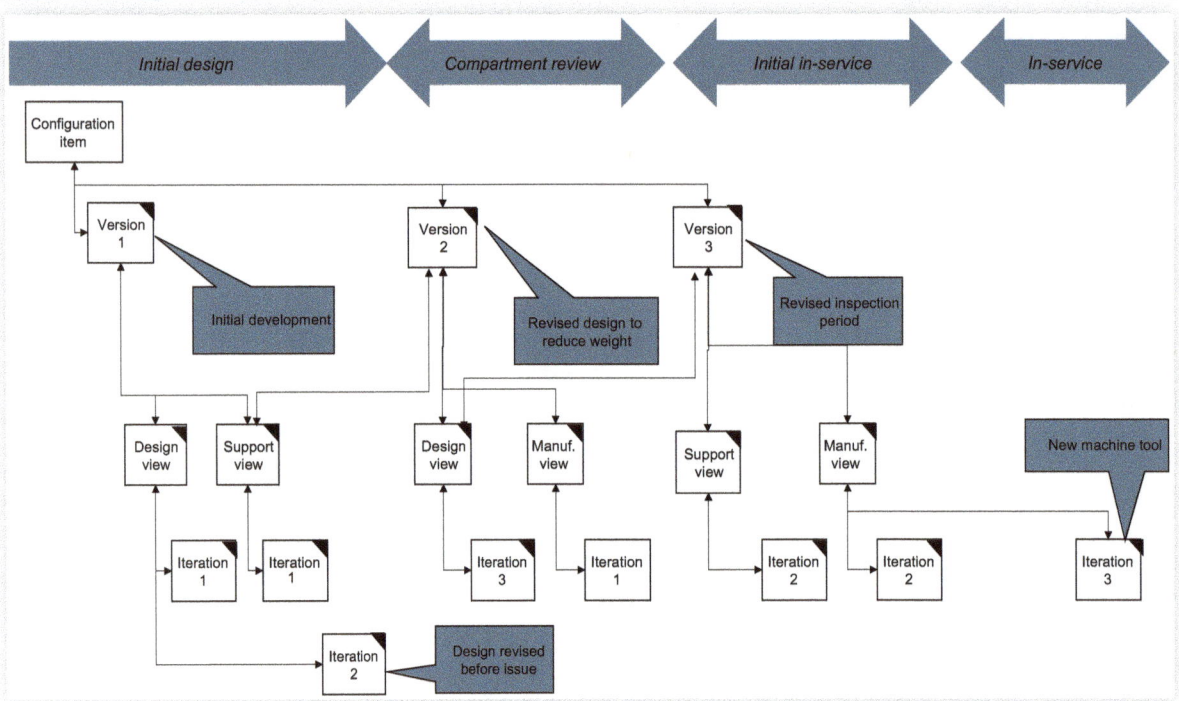

model. That process allowed for products derived from CAD, such as a drawing or a CNC part program, and these could be bundled with the CAD model without explicitly representing the view entity. PDM systems of the period did not include the view construct, and in STEP-based exchanges they inserted (and ignored) dummy view entities.[2]

Adoption of the view iteration counter would require an extension of the STEP standard and an update to configuration processes. It is likely that its absence will be fudged until there is substantial agreement on the use of IDEs and views. One way of fudging the issue is to use a sub-version numbering convention, in which minor issues (e.g., V1a to V1b) are for the release of views other than that of design, and major issues (e.g., V1b to V2a) are releases of a new design view.

11.3.3 Implications for Archiving

First, the IDE approach reinforces the value of submitting the models and the CI as separate items, since one would not want to rearchive the CAD model if a support procedure is updated. Second, if separate iterations of a view are signed off, then each view will have its own specific authorization and other process metadata. Here, instead of the CI being the target of a single SIP, there would be separate SIPs for the CI and each view.

Would it be possible to read an older CI without iterators into a new generation IDE? Since it is only the iterator entity that is missing, a dummy iterator entity could be inserted between the views and the models and it requires no new information.

---

[2] When this was pointed out, a prominent vendor admitted omitting the view entity, but then claimed that "no-one would want it." They no longer make that PDM system. Private correspondence c. 1996.

11.3.4 **Models as Databases**

One alternative line of software development is the multiview development application, such as seen, for example, in model-based systems engineering. Such software works using database objects corresponding to a real-world item such as an individual requirement or an aircraft system. A user will create a model within a particular view, either creating objects for their specific model or, more often, reusing objects created in different views. For example, a system model may have a structural view, a behavioral view, a view linking its timings to that of other systems, and so on. From a process perspective, this is another way of creating a team design environment. From a data model perspective, this is very similar to the IDE data model, but with finer-grained system objects replacing the separate models.

Although it is possible to archive the content of such a system in a single submission, like PDM, this system is a database in continuous evolution. A similar approach to PDM could be taken, dividing the database into submission objects based on their lifecycle, allowing each submission to reference objects in other submissions, and using a method such as promissory usage to allow the submission to validate that the targets for the links exist even though the item is not yet submitted.

11.4 **Summary**

The discussion of archiving PDM data in Chapter 10 got stuck because of the difficulties of creating references to items that had yet to be archived. There are basically two ways to resolve this. The first option is to accept that some of the items referenced will not be found in the archive and hope they are submitted at a later date. The alternative is to archive a "promissory" item, which can be used to verify the existence of the item and be the anchor for the link within the archive. Periodic reviews will check whether the promise has been fulfilled.

The development of integrated design environments will likely make the current model of configuration management obsolete. As well as being the context for the design view, the CI will be the context for manufacturing and support models, and they have a different lifecycle to design. It is proposed here to use the existing view entity together with a new view iteration counter to group the different perspectives on the CI. A configuration change follows from a change to the design view, but not from a change to the other views.

Is this enough to archive an aircraft design? What LOTAR gives is a paradigm of how to archive any independent design model. With the contribution here, PDM archiving gives a route to archiving the glue that sticks all the models into a coherent whole.

References

11.1 Wikipedia, "Common Object Request Broker Architecture," https://en.wikipedia.org/wiki/Common_Object_Request_Broker_Architecture, accessed August 8, 2019.

11.2 Wikipedia, "Situational awareness," https://en.wikipedia.org/wiki/Situation_awareness, accessed August 7, 2019.

11.3 Barker, S., "Extending STEP Support to Concurrent Engineering Development," presented at *the PDT Europe 2001*, Brussels.

Summary and Future Directions

Bacon is red, Some stuff is blue, Poems are hard, Eat here

—The wall of a bar in Reykjavik.

12.1 The Story So Far

If you have read this far, you will have absorbed principles culled from a couple of thousand pages of reports, with passing references to tens of thousands of pages of detail, and hints that this is just the tip of the iceberg. Engineers, researchers, and librarians have invested a great deal of time and effort into archiving, and the good news is that most of their results are available for free. The bad news is that to go any further, you will have to read some of that detail.

One key document is the OAIS recommendations (Chapter 3), which set out the concepts and the terminology for an archival system. An archival system is developed for a particular community—its designated community—and it works to ensuring that the data it holds remains understandable by that community for as long as it needs it. Importantly, the archival system is not just a collection of IT components, but a system that includes processes and an organization, the latter including archiving specialists. One specialist role is to watch for changes in technology and when it warrants, to refresh the storage technology or reformat the data. A second role is to understand the knowledge base of the designated community, and when that changes, find material to fill the gap between what they know now and what the original data creators thought they would know. Perhaps such first specialist was the mason who carved the Rosetta Stone [12.1].

Two further key documents come from the LOTAR project. LOTAR part 2 identifies the aerospace requirements (Chapter 2), particularly section 6 on key requirements. The second is part 3, on general principles. The central principle is that LOTAR does not aim to preserve the past by keeping old software running, but to sustain the design by making it accessible to the users in their day-to-day IT environment. To avoid loss of information, LOTAR identifies data standards, *key characteristics* that must be preserved, and *validation properties* that check they are preserved.

LOTAR has heavily drawn on the STEP standards (the ISO 10303 series) as a vendor-independent way of recording design data. These are long, highly detailed technical documents, which require knowledge of both business process and data modeling. For archiving, it is more important to know about them to understand them, although the most important experience is the pain of using less well-founded data models.

And finally, governance and planning (Chapter 6) identified who in a company needs to know what, and what decisions they need to make. All the rest is detail, and governance identifies who needs to know and act on that detail. To give an insight into that detail, CAD (Chapters 7 and 8) illustrates how to archive design models. Then PDM archiving (Chapters 10 and 11) illustrated the more complex world of business process data and the cross-linkages between models and processes.

12.2 **Going forward with LOTAR**

"How do you eat an elephant? One mouthful at a time." What this pleasantry ignores is that you also need a huge freezer, otherwise the meat will spoil long before you have eaten half a leg. Unfortunately, there is no freezer for an aircraft design. Equally unfortunately, there is no substitute for the careful, painstaking work of the LOTAR project. As Chapter 9 illustrates, they are working on many of the major design disciplines, while avoiding duplicating the work done in other sectors such as libraries.

In focusing on the results of LOTAR, it is easy to forget the path to get there. Industrial data standards such as STEP took a long time to establish, but once established they have proved extremely valuable. They have improved the quality of design information and substantially reduced the costs of collaboration, and both these features also have value for long-term archiving. However, having proved the value of the basic standards, it has been necessary to grow the scope of the standards so that they cover all the design data that needs archiving. Developing STEP AP 242 has been a major investment in time and effort. This was not simply a matter of getting the tech geeks to draw a few more diagrams, but a community-wide effort that needed to convince CAD software developers, specialist translator companies, and end user businesses that this was worth investing in and would advance their capabilities, not make them obsolete. And LOTAR has done sterling work in making this happen.

Another aspect of the work is developing validation properties. LOTAR has involved specialists in design-support software, commissioning them to identify potential properties and prove them through prototypes. These properties are then fed into the standardization process so that validation properties are stored with the models. The software vendors have been cajoled into implementing them, and encouraged to bring them forward into the various implementor forums (e.g., CAx-IF or PDM-IF).

And finally, LOTAR have engaged with the FAA and EASA, the US and European certification authorities, since they will ultimately judge whether the solutions proposed are acceptable for aircraft certification.

12.3 **What about XML, OWL, and so on?**

If you were to ring a number at random in China or Ethiopia, your main concern would be whether you could understand what the person on the other end was saying. You would not care about whether the line used VOIP or 5G technology, even if you know what they are. These are for the telephone companies' techies to sort out.

There are a number of IT technologies, which sections of the IT world have promoted over the last few years, including XML, RDF, or OWL (all have Wikipedia entries). These are concerned primarily with the technology of data representation—the IT techie domain. "The W3C Web Ontology Language (OWL) is a Semantic Web language designed to represent rich and complex knowledge about things, groups of things, and relations between things" [12.2] over-eggs what OWL does—the semantics referred to are those of a small group of predicates in formal logic. An OWL-based system still needs an information model to relate the terms of a semantic web predicate to business processes in the real world.

It is not that these technologies have no significance for an archive. However, the users of an archive are information providers and information users—they are concerned about what the information means to the business processes they operate, not how they get written in a computer file. Leave these technologies in the care of technical specialists who can advise on how the formats change over time (Chapter 6). And user organizations do need their own technical specialists to ensure that any archival system meets their actual needs, rather than the assumptions of an IT system's supplier. Remember that the problem of long-term archiving arises because the lifecycle of IT products is short compared to that of an aircraft.

12.4 **No Crystal Ball**

A little over 40 years ago, most computers were mainframes, sealed in air-conditioned rooms, and served by a priestly caste of systems programmers and computer operators. The main storage medium was half-inch magnetic tape. The first home computer appeared in the United Kingdom in 1977—the NASCOM 1, with 1 Kbyte of storage [12.3]—followed by the original PC from IBM in 1981, with 5¼ inch floppies [12.4]. The best guess is that computers will look different again in another 40 years.

The basics of geometry were notably laid out by Euclid in about 300 BC [12.5], although much of the mathematics was known earlier. That geometry is still used today in modern CAD systems, although the ways of representing the mathematics have changed over the centuries, most notably with Descartes (died 1650) [12.6]. In 40 years, geometry will look exactly the same.

The IGES standard for exchanging geometry dates from 1980 [12.7]. IGES was superseded by STEP (Section 7.3), which was identified as a standard for long-term archiving as early as 1995 [5.2]. Over the past 25 years, STEP has extended its scope while trying—and mostly succeeding—to maintain upward compatibility, so new versions of the standard do not break old data. If STEP is replaced over the next 40 years, the changes are likely to be mostly in the area of business process-defined data. For the mathematically based data definitions such as geometry, there will be, very likely, a directly computable path from STEP to any replacement.

Such forecasting is part of the *preservation planning* process for the archive. The hardest changes for an archival system to predict will be changes to the business processes arising from new aircraft technology. Forty years on from the Boys' Hobbies Annual cited in Chapter 1, valves in radios had been mostly replaced by semiconductors. How technology watch gets integrated with preservation planning will be one of the challenges for the longer term.

12.5 Summary of a Summary

Digital archiving might initially seem straightforward—saving data—but the deeper one goes, the more complications appear. Saving data only works in the short term-3-5 years, say. But then the software changes, the pictures decide that they will only appear in the header on the next page and equations appear as a line of black squares.

That's when you discover you want to save information—not the "I" of IT, but the very stuff of business process. You need to get under the skin of the business process to understand the detail below.

But what about the long term? Then you need to get into knowledge. To simplify the problem, you will set the boundaries of a *designated community*, the people who will need to know how to read the information you have saved. And then you will need to work out what knowledge your data creators are using when they create designs. This is the point where archiving theory is yet to catch up, and a journey for the future:

> *Whan that Aprill with his shoures soote*
>
> *The droghte of March hath perced to the roote ... [12.8]*

Notes and References

12.1 Wikipedia, "Rosetta Stone," https://en.wikipedia.org/wiki/Rosetta_Stone, accessed July 2, 2019.

12.2 W3C, "Semantic Web OWL," https://www.w3.org/OWL/, accessed July 1, 2019.

12.3 Wikipedia, "NASCON (Computer kit)," https://en.wikipedia.org/wiki/Nascom_(computer_kit), accessed July 1, 2019.

12.4 Wikipedia, "IBM Personal Computer," https://en.wikipedia.org/wiki/IBM_Personal_Computer, accessed July 1, 2019.

12.5 Wikipedia, "Euclid," https://en.wikipedia.org/wiki/Euclid, accessed July 1, 2019.

12.6 Wikipedia, "Rene Descartes," https://en.wikipedia.org/wiki/Ren%C3%A9_Descartes, accessed July 1, 2019.

12.7 Wikipedia, "IGES," https://en.wikipedia.org/wiki/IGES accessed July 1, 2019.

12.8 Winny, J., Ed., *The General Prologue to the Canterbury Tales* (Cambridge: Cambridge University Press, 1965).

glossary

Access agreement - Agreement between the consumer and the archive allowing access to the archive

Additive manufacture - 3D printing or similar technique used in manufacture

AECMA - European Association of Aerospace Industries

AIA - Aerospace Industries Association

AIP - Archival Information Package

Airworthiness - Demonstration that an aircraft or its component is fit to fly

Airworthiness limitation - Restriction on flight operations or maintenance schedules

AP 203 - Configuration controlled 3D design of mechanical parts and assemblies

AP 214 - Core data for automotive mechanical design processes

AP 239 - Product Lifecycle Support

AP 242 - Managed model-based 3D engineering

Application Activity Model (AAM) - Process model giving the context for a STEP AP

Application Interpreted Construct (AIC) - Reusable STEP component

Application Interpreted Model (AIM) - STEP data structure used in exchange

Application Protocol (AP) - STEP information model fulfilling a business requirement for data exchange

Application Reference model (ARM) - business process model for a STEP AP

Archiving - Long-term storage of information—connotations of file and forget (see Figure 1.4)

ASD - Aerospace and Defence Industries of Europe

ASD-STAN - Standardization arm of ASD

ASME - American Society of Mechanical Engineers

ATA - Air Transport Association

Biarc - A 2D curve composed of pairs of circular arcs

Bicubic spline - A 2D curve composed of segments defined by cubic equations

Bill of Materials (BOM) - A list of parts in an assembly

Boundary Representation (Brep) - Type of 3D model, *see* Surface modeler

Business process data - Metadata used as part of a business process, *see* MUST of metadata

CAD - Computer Aided Design

Certification - Formal approval by an appropriate authority that something is fit for purpose

CNC - Computer Numerically Controlled (machine tool)

Composite structure - Component made from a composite material, such as carbon fiber

Computational tolerance - The errors that result from numerical calculations because of the way numbers are approximated in a computer

Computer Aided X-Implementors Forum (CAx-IF; X may be Design, Manufacture, etc.) - Test CAD standards and data exchange interfaces

Configuration control - Formal control of the components and build of an aircraft

Configuration Item (CI) - Something identified as requiring configuration control

Configuration item - Something identified in a configuration plan as requiring configuration control

Configuration Management (CM) - The overall process covering change control, configuration identification, status accounting, and audit

Conformance Class - A subset of a STEP AP used for exchange between particular tool sets

Constructive Solid Geometry (CSG) - A type of 3D model which represents solids directly

Content information - The target of preservation comprising a content object and its representation information

Context information - Information relating the Content Information to its environment, including other Content Information

Co-operating archives - Archives sharing SIPs and DIPs allowing them to ingest items from each other

Core model - The information model needed to represent the essential information for an archival use case

Cost Model - A model of the cost of operating a system or services

Data curation - Term used in the libraries and academic communities covering archiving

Data management - The management of physical computer files containing data

Data - The representation of a fact as a number or text string

Derived format - A translation of software's native data format, usually to a form more suitable for archiving

Descriptive Information - Information used by an OAIS data management item to support finding, ordering, and retrieving Content Information

Design Authority - The organization with legal authority (and liability) defining a design

Design intent - The engineering significance of a design element

Designated community - The community of information users who should be able to understand a particular set of Content Objects

Digital Signature - A method of ensuring that a set of information cannot be tampered with, possibly implying some level of approval

Dissemination Information Package (DIP) - In OAIS, the information delivered to a consumer to fulfill an access request

Drawing-based - Where a drawing is the authoritative source for a design

EASA - European Aviation Safety Authority

Effectivity - A method of associating the usage of a part with the version of the products it is used in

EIA - Electronic Industries Alliance

Essential Information - The information needed to represent the design intent of a model

Evidential weight - A legal term for proof of the provenance and authenticity of information used as legal evidence

Exact (solid, surface) - A model that describes the exact shape of something, contrasted with "tessellated" or "faceted"

EXPRESS - An information modeling language used in STEP and other information standards

FAA - Federal Aviation Authority

Faceted model - A model that approximated the shape of something, contrasted with "exact"

Federated archives - A group of archives making content accessible by common finding aids

Finite element - A method used to calculate the properties of something by breaking it into many simple elements

Fit, form, function - Any change to fit, form, or function requires a formal configuration change authorized by a Work Order

Fixity information - Ensures that the Content Information is unaltered in the archive

Hearsay - Evidence, such as a computer printout, not directly testified in court, *see* Evidential weight

IDE - Integrated Design Environment

Information - The mapping between a situation and the data that represents it—what a data element refers to

Ingest - The process of adding information to an archive

ISO 10303 - The series of standards referred to as STEP

ISO - International Organization for Standards

IVHM - Integrated Vehicle Health Monitoring

Key characteristics - The features of a material or part whose variation has a significant influence on fit, form, function or service life

Knowledge - Information used intelligently—the context that defines the meaning of information

Long-term - A period crossing more than one generation (major rewrite) of software

LOTAR - LOng Term Archiving and Retrieval of digital technical product documentation such as 3D CAD and PDM

Manufacturing tolerance - The variation on the shape permissible for a manufactured item

Medium-term - A period crossing no more than one generation (major rewrite) of software

Metadata - Literally, data about data—general class for which only the subclasses have any specific meaning

Model-based - Where a computer model is the authoritative source for a design

MUST of metadata - Major classes of business process data—management, usage control, search, and trust

NAFEMS - International Association for the Engineering Modelling, Analysis and Simulation

Native format - The default storage format peculiar to a particular software package

NATO stock number - A part type identification code controlled by NATO, which requires all parts with the same code to be interchangeable

NURB - Nonuniform rational B spline

OAIS - Open Archival Information System

OASIS - Organization for the Advancement of Structured Information Standards (search OASIS-open)

OWL - Web Ontology Language

Packaging information - Information used to describe and decode an OAIS information package

Part number - A unique number identifying a part, usually allocated through the configuration management system

PAS - Publicly available specification

PDES Inc - US-based consortium supporting model-based engineering, and supporting LOTAR

PDF - Portable Document Format

PDM Schema - A subset of a STEP developed for exchanging PDM data

PDM - Product Data Management

PLCS - Product Life Cycle Support

PMI - Product & Manufacturing Information

PREMIS - Preservation Metadata: Implementation Strategies

Presentation - A rendering of a model, such as a drawing or screen shot

Preservation description information - Information necessary for the adequate preservation of a Content Object

Preservation planning - Planning to ensure the designated community can read the content of the archive in the future

Producer - The creator of information destined for an OAIS

Product Data Management-Implementors Forum (PDM-IF) - Tests PDM standards and data exchange interfaces

Product structure - The hierarchical breakdown of a product into its components, according to a view of the product

ProSTEP iViP - European-based organization supporting product data management and virtual product creation

Provenance - A chain of evidence and data identifying the origins of data and its passage to the archive

RASSC - Retention and Access Services in Supply Chains

Reference information - Information that identifies a Content Object and its contents

Representation information - Describing how a content object is structured

Representation network - A network of representation information, arising from the need to record how representation information is itself represented

Representation - In STEP, the information structure representing a property of a model, such as shape

Requirements engineering - The engineering process for forming complete, unambiguous, and verifiable requirements

Resource Description Framework (RDF) - A semantic web construct

Retention - Long-term storage of information—connotations of keep available (see Figure 1.4)

SASIG - Strategic Automotive Product Data Standards Industry Group

Service stack - A set of services in which higher-level services reuse lower-level services

Service - In software, where a function is provided for a subscription to a service provider

SGML - Standard Generalised Markup Language, describing how a document is structured

Short-term - Occurring within a single software generation

SIP - Submission Information Package

SLA - Service Level Agreement

Software generation - A major software rewrite, often involving substantial changes to the internal data model

Solid modeler - A CAD modeler in which a solid is represented directly, as opposed to through its surfaces

Standard format - Data stored using a standard data model, usually one which is public and under the control of an independent organization

STEP - Standard for the Exchange of product Model Data (ISO 10303)

Strict liability - (legal) Liability that does not depend on actual negligence or intent to harm

Submission agreement - Agreement between the producer and the OAIS on the contents, format, and process for submission

Surface modeler - A CAD modeler representing a solid as a set of connected surfaces

Sustainment - Long-term storage of information—connotations of keep usable (see Figure 1.4)

SysML - Systems engineering modeling language

Systems engineering - An engineering discipline based on systems thinking—focused on properties than form

Tangent continuity - A constraint that ensures curves and surfaces are smooth

TDR - Trusted Digital Repositories (ISO 16363)

Technical metadata - Data relating to the structure of the target data

Tessellation - The division of a surface into tiles

Tolerance - The numerical errors that can accumulate in a geometric calculation

Tolerance - The variation on the shape permissible for a manufactured item

TRAC - Trustworthy Repositories Audit & Certification

Type certificate - Certificate from a regulatory body showing the airworthiness of an aircraft type

Type design - The design of a type of aircraft

Use case - A situation where an artifact could be used

Validation property - Information calculated from the original model used for validation

Validation - Checking that, when recovered into modeling software, the key characteristics of the model have not changed

VDA - Germany Association of the automotive Industry

Verification - Checking the quality of a model based on its representation

Visualization - A model, typically simplified, used to present a design to a user

Wire frame - A CAD modeler in which a solid is represented only by its edges

Wiring loom - A collection of wires installed as a single item

Work order - A authorization specifying the Configuration Items to change and what changes to make

Work request - A request to change a design, linking the requirements to the Work Orders

XML - eXtensible Mark-up language, used in specifying data formats